PROCEEDINGS

OF THE

NATIONAL

SHIP-CANAL CONVENTION,

HELD AT THE CITY OF CHICAGO,

JUNE 2 AND 3, 1863.

CHICAGO:
TRIBUNE COMPANY'S BOOK AND JOB PRINTING OFFICE, 51 CLARK STREET.

1863.

R

PROCEEDINGS.

CALL FOR A NATIONAL SHIP-CANAL CONVENTION.

WASHINGTON, D. C., March 2, 1863.

Regarding the enlargement of the Canals between the Valley of the Mississippi and the Atlantic as of great National, Commercial and Military importance, and as tending to promote the Development, Prosperity and Unity of our WHOLE COUNTRY, we invite a meeting of all those interested in the subject, in Chicago, on the first Tuesday in June next. We especially ask the co-operation and aid of the Boards of Trade, Chambers of Commerce, Agricultural Societies, and Business Associations of the country.

Edward Bates, Attorney General U. S.

Members of the House.

Isaac N. Arnold, Illinois.
A. G. Riddle, Ohio.
E. B. Washburne, Illinois.
H. L. Dawes, Massachusetts.
A. B. Olin, New York.
Justin S. Morrill, Vermont.
E. G. Spaulding, New York.
S. Hooper, Massachusetts.
Portus Baxter, Vermont.
Schuyler Colfax, Indiana.
George P. Fisher, Delaware.
Augustus Frank, New York.
Cyrus Aldrich, Minnesota.
R. E. Trowbridge, Michigan.
Samuel L. Casey, Kentucky.
Jos. Segar, Virginia.
W. D. McIndoe, Wisconsin.
F. C. Beaman, Michigan.
W. P. Sheffield, Rhode Island.
Alfred Ely, New York.
J. M. Ashley, Ohio.
Gilman Marston, New Hampshire.
F. F. Low, California.
Samuel T. Worcester, Ohio.
John W. Wallace, Pennsylvania.
Benjamin F. Thomas, Massachusetts.
T. C. Phelps, California.
Thomas D. Eliot, Massachusetts.
William J. Allen, Illinois.
A. A. Sargent, California.
W. E. Lansing, New York.
Jesse O. Norton, Illinois.
P. B. Fouke, Illinois.
George W. Julian, Indiana.
W. R. Morrison, Illinois.
Wm. Morris Davis, Pennsylvania.
William Kellogg, Illinois.
J. M. Goodwin, Maine.
Stephen Baker, New York.
James S. Rollins, Missouri.
C. W. Dunlap, Kentucky.
Thomas L. Price, Missouri.
J. C. Robinson, Illinois.
Horace Maynard, Tennessee.
Charles Delano, Massachusetts.
F. W. Kellogg, Michigan.
A. J. Clements, Tennessee.
John H. Rice, Maine.
S. N. Sherman, New York.
A. W. Clarke, New York.
Theodore M. Pomeroy, New York.
R. E. Fenton, New York.
A. S. Diven, New York.
Burt Van Horn, New York.
R. B. Van Valkenburg, New York.
M. F. Conway, Kansas.
Wm. Windom, Minnesota.
Dwight Loomis, Connecticut.
R. Franchot, New York.
C. H. Van Wyck, New York.
Elijah Ward, New York.
John F. Potter, Wisconsin.
James B. McKean, New York.
Wm. Vandever, Iowa.
Owen Lovejoy, Illinois.

E. P. Walton, Vermont.
John Hutchins, Ohio.
W. H. Wallace, Washington Territory.
Edward Haight, New York.
A. L. Knapp, Illinois.
George C. Woodruff, Connecticut.
Amasa Walker, Massachusetts.
B. F. Granger, Michigan.
Edward H. Smith, New York.
John B. Alley, Massachusetts.
A. S. White, Indiana.
Samuel C. Fessenden, Maine.
S. Edgerton, Ohio.
James H. Campbell, Pennsylvania.
H. P. Bennett, Colorado.

Members of the Senate.
J. R. Doolittle, Wisconsin.
James Harlan, Iowa.
James Dixon, Connecticut.
H. M. Rice, Minnesota.
James A. McDougall, California.
J. B. Henderson, Missouri.
J. H. Lane, Kansas.
R. Wilson, Missouri.
S. C. Pomeroy, Kansas.
L. Trumbull, Illinois.
W. A. Richardson, Illinois.
Charles Sumner, Massachusetts.
Henry Wilson, Massachusetts.
J. M. Howard, Michigan.

PURSUANT to the above call, delegates from nearly all the loyal States assembled at Chicago, on June 2, 1863, to the number of about five thousand. The day was propitious; and a spacious tent was erected for the accommodation of the delegates, on Lake Park, between Eldridge and Harmon Courts, and overlooking the blue waters of Lake Michigan.

MORNING SESSION.

The Convention was called to order at eleven o'clock A.M., by Dr. D. BRAINARD, of Chicago, who nominated the Hon. CHAUNCEY FILLEY, Mayor of St. Louis, for temporary Chairman.

The nomination was agreed to, and, upon his introduction to the Convention by Dr. BRAINARD, the CHAIRMAN said:

Gentlemen of the National Convention:—I thank you for the honor you confer upon the State of Missouri, and the city of St. Louis. I have been sent here by the people of that great commercial metropolis, to represent her interests in this meeting. I can give to you the objects and views of the City Council of St. Louis, with reference to the business before us, by reading the preamble and resolutions which they have passed on the subject, better, perhaps, than I could do in any other way. They are to the following effect:

WHEREAS, A National Convention is about to assemble at the city of Chicago, to deliberate in reference to the proposed enlargement of the Illinois and Michigan, and New York and Erie canals; therefore,

Be it Resolved, By the Common Council of the city of St. Louis, that, as the representatives of this great commercial metropolis, we look with much interest and anxiety to the accomplishment of the project proposed for the consideration of the Convention.

And be it further Resolved, That the President of this Council be, and he is hereby, instructed to appoint five members thereof, to represent the city in her corporate capacity in the said Convention; and that his Honor, the Mayor, be requested to act in conjunction with the delegates so appointed; and that the Mayor be authorized to appoint five delegates outside of said Council, to represent the city in said Convention.

Dr. BRAINARD. Mr. Chairman, I would nominate for Corresponding Secretaries, Hon. MOSES M. STRONG, of Wisconsin; A. M. CLAPP, Esq., of Buffalo; and W. H. MCHENRY, Esq., of St. Louis.

The nominations were confirmed.

The CHAIRMAN introduced the Rev. BISHOP SIMPSON of the Methodist Episcopal Church, who, amid reverential silence, came forward, and invoked the Divine blessing as follows:

PRAYER.

Almighty and everlasting God, our Heavenly Father; Thou art the Creator of all Worlds, the Sovereign of all Nations; All wisdom is from thee; Thou rulest in the armies of Heaven, and on the Earth beneath. Assembled as we are this day from various States, Districts and Cities,—in commencing our deliberations on matters which may be of vast importance to our Country, and of untold interest to millions of our Posterity, we desire to invoke thy presence, and thy blessing; for we know, that without thy aid, no good can be accomplished, and that the hearts of the children of men are in thy hands. We adore thee for all thou hast done for us; for the goodly land which thou hast given us; for its mountains and its plains; for its lakes and its rivers; for its genial clime and its fertile soil. We bless thee for those rays of Science, and those movements in Art, which thou hast sent abroad upon the Earth, and which have developed in part the resources which thou hast given us. We bless thee that our mountains have been pierced, so that our lands might be girt by bands of iron. We praise thee for the diffusion of Intelligence, and for the blessing of a high order of Christian Civilization; and we pray that these blessings may continue to abound among us, until all parts of the Country shall be united indissolubly together; that Art and Science shall bless the Land; until all the resources of this vast country shall be developed under thy smile; and that the light of thy countenance shall be and remain upon this people.

And, O Lord; we pray that thou mayest be with these Delegations; may no sickness come upon them; may no accident befall them; may they be returned to their families in safety; and may there be cordial greetings, without one pang of woe. May thy blessing so rest upon them in all their deliberations, that, in the final issue, they may be found to have consulted together wisely and harmoniously; and may they see the issue happily consummated, and the whole Country and People more closely united than ever before.

Hear us for our Country! God of our Fathers; Look upon our Nation in love. Bless thy servant, the President of the United

States; may the Spirit of Wisdom be upon him, and all who are associated with him in Cabinet council. Bless the Governors of States, and the Members of Congress; may they devise such laws and measures, as shall be for the Interest, Peace, and Prosperity of our Land. Guide all our officers, civil and military, to such measures as shall be for the benefit of our great and wide-spread Nation. We pray thee, O Lord, to let this terrible conflict very speedily come to a close. If it be thy will, Lord, may Peace take the place of War; and, on the principles of Equity and Righteousness, may our Nation be united and happy;—a prosperous Nation, whose God is the Lord. May the Voice of War be heard amongst us no more, and the Spirit of Peace and Consolation be spread over the Earth, until the family of Nations shall dwell together as one vast and universal Brotherhood.

Hear us, O Lord, while we further pray, in the language which thou hast taught us:

Our Father who art in Heaven, Hallowed be thy Name; Thy Kingdom come; Thy Will be done on Earth as it is in Heaven; Give us this day our daily bread; And forgive us our trespasses, as we forgive those who trespass against us; And lead us not into temptation; But deliver us from evil: For thine is the Kingdom, and the Power, and the Glory, forever and ever. Amen.

The CHAIRMAN. Dr. BRAINARD will now welcome the delegates on the part of the city of Chicago.

Dr. BRAINARD then addressed the Convention as follows:

Mr. Chairman, and Gentlemen of the Convention:—I appear before you in behalf of the Board of Trade of Chicago, the Mercantile Association, and the citizens in general, to bid you welcome to this City, and the West. Chicago, although not possessing the historical associations or venerable institutions which give interest to the cities of older States, presents an example of what may be done by unaided industry, and exhibits the spectacle of an intelligent and free people establishing by themselves the foundations of society. She already embraces within her limits the largest granaries of the world, and numerous important manufactures; but, more than all, she has a population unsurpassed in enterprise and patriotism, who unite, one and all, in offering you a heart-felt welcome.

Your presence here fills us with joy. To see so many bearing honorable names; so many representatives of other States and large cities here assembled; fills us with gratitude which no words can express.

The occasion which has called you together is one of no ordinary character. It is not the call of a famishing people, nor of cities threatened by hostile armies; it is the voice of men shut out from the markets of the world, oppressed by the excessive productions of their own toil, which remain wasting and worthless upon their hands, depriving labor of half its rewards, discouraging industry, and paralyzing enterprise. In their distress they called upon the National Legislature—that government to which they owe allegiance, and to the support of which their lives and property are devoted—and failed to obtain that relief which they had a right to expect. Now they appeal to the people themselves—to the wise, the enlightened, the patriotic, and the powerful—in a word, they appeal to you, gentlemen, and ask you to speak in their behalf that word of power which, in a good cause and in a free government, will not be spoken in vain.

Need I tell you of the history of this region in whose behalf your sympathies are asked? This city, Illinois, and her sister States of the North-West, are but of yesterday. Many of you have witnessed their birth and their growth; with their present population and its increase, their extent and fertility, their commerce and productions, and their capacities for boundless expansion in the future,—all must be familiar; and to dwell upon them would be to repeat a tale already many times told.

To others more capable has, moreover, been committed the task of supplying those facts, by which a certain judgment may be formed, and on which intelligent action may be based.

The sons of the West have demonstrated their loyalty, their patriotism, and their courage, on every battle-field. The pages of history on which their actions are written, are not dimmed by a single blot. Their names are inscribed on the "roll of honor," along with those which time shall not efface, but gild with brighter lustre through coming centuries. Even now the southern sky is lit up by the glory of their deeds, and the nation looks to their action for its preservation, with a hope that will never be disappointed. Such are the men for whom your efforts are invoked.

Our fertile prairies lay for centuries untouched by the hand of agriculture; not because they were unknown, but because they were inaccessible. Their prosperity dates from the opening of the Erie canal, thirty-eight years ago. The increase of the West, and its productiveness, have all been the direct result and consequence of the construction of that and other channels of trade and travel. To the genius of De Witt Clinton, these States owe their existence, and the Nation its present strength and greatness. Every city here is a monument to his honor. From every domestic altar, where morning orisons, or evening thanksgivings are offered to the Giver of all Good for unknown blessings, do benedictions upon his name ascend. The policy which he established has stood the test of time, and received the seal of success; but its results have so far exceeded his expectations that the channels he projected—enlarged and multiplied as they have been—are all filled to their utmost capacity. The increase of the West may be stopped, her fertile fields deserted for the mines of Oregon, California, and Colorado; her bright future be darkened, and her people discouraged, by the refusal of the Government to open those means of communication, upon which her growth and prosperity depend. Her commerce begins to feel the chain which fetters it; her people already complain that the fruits of their labor are gathered up by others; and, while carriers are enriched, consumers and producers suffer alike, and are impoverished. Under these circumstances, she asks of Congress to construct from the Mississippi river to the Atlantic a channel, adequate to the carrying of her staples, and supplying her wants, without unnecessary delay or exorbitant charges.

For our own part of the work, from lake Michigan to the Mississippi river, there will be transmitted to you surveys, plans and estimates, showing its cost, practicability and usefulness, to which we invite the rigid scrutiny of engineers and business men. For the Eastern division, without presuming to designate the route, what we would ask is, that it should be broad and deep; and, if practicable, that there should be more than one; and the history of the past justifies us in the promise, that, however capacious they may be, wherever situated, by whom constructed, they will all be speedily filled to overflowing.

The commerce of the West has enriched the country through which it has passed. It has raised New York to a degree of wealth and credit, which rivals that of the most powerful kingdoms. It has preserved and increased the exports and the revenues of the nation in a period of war, when other sources were cut off. It is yet in its infancy. Who shall dare to arrest its progress? Who shall stand between the West and the hungry nations of the Old World? Whoever he may be, it is safe to predict that he will be swept away, consigned to infamy, or buried in oblivion.

And what objection can be urged against a project so feasible and so necessary? Is it that it is commercial and not military in its character? Look at these shores—at this lake covered with rich cargoes;—not a gun, not a ship for their defense. Yet they are accessible for armed vessels from Plymouth and Portsmouth; and the hands of our Government are tied by treaty so that no navy can be built upon these shores, or launched upon these waters. Nowhere in the world is commerce so valuable, so left without protection. It is essential to our national existence, and so long as it remains without means of defense, England will be able to exact her own terms in all disputes with our Government, by an exhibition of force which it has no means of resisting.

It is necessary to provide for a naval contest on these Lakes, and in no way can this be done so readily and effectually as by connecting them with the Atlantic Ocean and Gulf of Mexico, by ship and steamboat canals. These, even if they were never to be used for the passage of armed vessels, would still be of the greatest service in case of war. By augmenting the national wealth, by attracting population, by facilitating transportation of food and stores, they would do more for our military power than the raising of armies, or the building of navies. But the spirit of unity and nationality, which the mere commencement of this work would engender, is above all other considerations. By blending into one the two great systems of inland navigation of the continent, it will cause the life-blood of commerce to move in a single current through all its channels, as in the human system the blood is impelled through all the arteries and veins by a single impulse of the heart. Thus, out of conflicting parts will a homogeneous nation be formed, with one spirit, and the same common interests; with a power which may be concentrated at every point, and an internal commerce which man cannot interrupt; such a nation never fears either foreign or domestic foes.

Gentlemen, the object for which you are assembled excels in magnitude, any ever offered for the consideration of a similar body. Not only will the work proposed join the Lakes with the Mississippi River and the Atlantic, but it will form part of a great highway from the Atlantic to the Pacific, by means of which the wealth of Asia on the one hand, and of Europe on the other, may be grasped and made to pass through the Bay of New York, and the Golden Gate of San Francisco; thus encircling the whole earth, and bringing all nations to pay tribute, and bow before the sceptre of our commerce.

It looks forward to coming time, beyond the scope of human vision. May it please God to enable you to comprehend and accomplish the work set before you. It will be the grandest step in human progress which the world has yet witnessed, over which religion, humanity and patriotism will rejoice, and for which the voice of future millions, unknown and unnumbered, shall load your memory with perpetual gratitude.

To accomplish objects so vast will require not only labor, perseverance and determination; but a spirit of justice, of moderation and conciliation, on the part of each and every member of this Convention.

There are several routes whose claims are to be consulted. Without presuming to indicate which of them may be most meritorious, I may say with truth, in behalf of this city, that she has no jealousy on this subject, and will be best pleased with the largest number of channels, which can promote the prosperity of the entire West; trusting to the enterprise of her merchants to secure her share of that general prosperity, on which the welfare of our city must depend. Such, gentlemen, is the spirit in which Chicago welcomes each member of this Convention.

Mr. SEDGWICK, of New York. In view of the permanent organization of this Convention, I beg leave to offer the following resolution:

Resolved, That a committee of two from each State, and one from each Territory represented, and also one from the District of Columbia, shall be appointed to report to this body the permanent officers of the Convention, consisting of a President, Vice President from each State, and five Secretaries, and also rules for the conduct and government of the Convention.

The resolution was agreed to.

On motion of Mr. KELLOGG, of Illinois, it was

Resolved, That, to enable the several delegations to name their respective members of the Committee on Organization, this Convention will take a recess of twenty minutes.

A delegate proposed to extend the time of the recess to an hour.

Mr. KELLOGG. I will suggest that the Illinois delegation have already agreed upon their men for the Committee on Organization. I have no objection to extend the term of the recess to one hour, if that time should be required; but the object in offering the resolution was to expedite the permanent organization; and it is important, in view of the business before us, if we take a recess at this time, to make it as short as possible.

A DELEGATE. Say thirty minutes, then.

Mr. KELLOGG. Thirty minutes was what I was about to suggest.

The resolution thus modified was adopted; and, accordingly, the Convention took a recess of half an hour.

At the expiration of the recess, the Convention was called to order by the Chairman; and the Secretary, on the part of the several State delegations, reported the following Committee on Nominations, required by the resolution of Mr. Sedgwick, to wit:

COMMITTEE ON NOMINATIONS.

Michigan,	Messrs.	E. G. Merrick, J. B. Crippen.
New York,	"	John A. King, Geo. S. Hazard.
Kansas,	"	A. M. Sawyer, L. Houston.
Missouri,	"	Wayman Crow, Chas. D. Drake.
Indiana,	"	Geo. W. Julian, W. W. Higgins.
Vermont,	"	S. B. Platt, Louis Follett.
New Jersey,	"	J. B. Sayer, Joseph C. Jackson.
Minnesota,	"	J. Z. Werst, R. A. Mott.
Rhode Island,	"	Comfort Tiffany, R. J. Arnold.
Massachusetts,	"	D. L. Harris, J. C. Converse.
Maine,	"	Israel Washburne, T. C. Hersey.
Connecticut,	"	John V. Elton, Thomas Porter.
District of Columbia,	"	W. H. Lamon.
Ohio,	"	R. P. Spaulding, Theodore Bissell.
Wisconsin,	"	H. L. Palmer, M. M. Davis.
Iowa,	"	W. I. Gilchrist, D. W. Kilburne.
Illinois,	"	I. N. Arnold, O. C. Skinner.

Mr. Kellogg. I am informed by Mr. Edwards, of New Hampshire, that he is not advised of any other delegate but himself from that State. I therefore move that the Hon. T. M. Edwards, of New Hampshire, be added to the Committee on Nominations.

The motion was agreed to, and Mr. Edwards was placed on the committee.

Mr. Foster, of Illinois, proposed the following resolution:

Resolved, That Governors and ex-Governors of States, Senators and ex-Senators of the United States, members and ex-members of the House of Representatives of the Congress, Presidents of Boards of Trade and Chambers of Commerce, Presidents of Agricultural Societies, Mayors of cities, Major and Brigadier Generals of the United States Army, and Canal Commissioners and ex-Canal Commissioners of the several States, be respectfully invited to take seats upon the platform, and that the Secretary be directed to issue tickets to them, and to the permanent officers of the Convention.

The resolution was adopted.

Mr. King, of New York. I understand that the names of the gentlemen forming the Committee have been announced. I take the liberty of proposing that they retire immediately after the adjournment of this Convention, and meet at the Sherman House, where there is a large room for their reception. I also make the motion that this Convention adjourn till three o'clock.

Mr. Partridge, of Missouri, seconded the motion.

Mr. Drake, of Missouri, suggested that, previous to going to the Sherman House, the members of the Committee come upon the platform, and make themselves acquainted with each other, and then proceed in a body to the Sherman House.

The motion was carried.

And the Convention then adjourned.

AFTERNOON SESSION.

The temporary Chairman called the Convention to order at three o'clock p. m., and announced the order of the report of the Committee on Organization.

Mr. King. The Committee appointed by the Convention to present the names of officers, beg leave to report:

FOR PRESIDENT.

Hon. Hannibal Hamlin.

FOR VICE PRESIDENT AT LARGE.

Gen. Hiram Walbridge, of New York.

FOR VICE PRESIDENTS.

Maine,	Mr.	Jedediah Jewett.
New Hampshire,	"	T. M. Edwards.
Vermont,	"	Lewis Follett.
Massachusetts,	"	Lorenzo Sabine.
Rhode Island,	"	R. J. Arnold.
Connecticut,	"	Calvin Day.
New York,	"	Erastus Corning.
New Jersey,	"	Marcus L. Ward.
Ohio,	"	M. D. Scott.
Indiana,	"	George W. Julian.
Illinois,	"	Joseph W. Singleton.
Michigan,	"	A. E. Bissell.
Wisconsin,	"	Levi Blossom.
Minnesota,	"	James M. Taylor.
Iowa,	"	Ebenezer Cook.
Missouri,	"	George Partridge.
Kansas,	"	Marcus J. Parrott.
District of Columbia,	"	Peter Parker.

FOR SECRETARY AT LARGE.

Mr. J. W. Foster, of Illinois.

FOR SECRETARIES.

Maine,	Mr.	Jonas H. Perley.
New Hampshire,	"	Wm. E. Chandler.
Vermont,	"	L. B. Platt.
Massachusetts,	"	D. L. Harris.
Rhode Island,	"	Comfort Tiffany.
Connecticut,	"	S. S. Gold.
New York,	"	S. M. Chamberlain.
New Jersey,	"	B. M. Price.
Ohio,	"	Mr. John E. Cary.
Indiana,	"	Mr. A. E. Drapier.
Illinois,	"	Chas. H. Lanphier.
Michigan,	"	Benj. Follett.
Wisconsin,	"	Moses M. Strong.
Minnesota,	"	R. A. Mott.
Iowa,	"	J. W. Van Orman.
Kansas,	"	D. W. Wilder.
District of Columbia,	"	F. A. Stringer.

Mr. KING continued. This, Mr. Chairman, is one part of the report of your Committee. I will submit the other part—that with reference to order of business and rules—after this shall have been disposed of.

The nominations of officers submitted by the Committee were then unanimously agreed to.

The temporary CHAIRMAN requested Brig. Gen. WILLIAM T. STRONG and the Hon. WILLIAM E. ARMSTRONG to conduct Mr. HAMLIN, the President elect of the Convention, to the chair.

SPEECH OF HON. HANNIBAL HAMLIN.

The Hon. HANNIBAL HAMLIN, delegate from the State of Maine, and Vice President of the United States, in taking the chair, spoke as follows:

Gentlemen of the Convention:—I tender to you my cordial thanks for this expression of your confidence and respect, in selecting me to preside over your deliberations. Relying upon your courtesy and your co-operation, I assume the responsibilities which you have devolved upon me.

You have met, Gentlemen, at a period of time remarkable in the history of our country. Obedient to the call of prominent citizens in different sections of the Union, you have come together, in this marvel of a city, that sits like a queen on the borders of this mighty Lake, to deliberate on questions that are almost as important as the existence of the Government under which you live. And there is one fact that challenges our observation, and justifies a remark. At a time when war with its red wave is deluging our land, and when brave, noble hearts, and strong arms have gone forth to defend the integrity of the common country, and to perpetuate the national existence, this vast concourse of people has assembled, representing all the varied interests of the broad land. That fact speaks a language far more significant than any that I can utter. It announces that we are here to prepare for the necessities of the future.

But a few years since, the people of this mighty, central portion of our country was numbered almost in units, and its commercial statistics found no place in commercial records. And what do we find to-day? Why, Gentlemen, you count your commerce by the half thousand of millions, and your population by some ten millions. The sagacity of that statesman of whom the Empire State may well be proud, and in whom every American citizen has also his pride, did effect the opening up of the then little commerce of the lakes—small, as compared with its present,—and you have come to-day to imitate his example, and to initate a policy, I trust, that shall secure those works that are commensurate with the demands of to-day. Let me tell you, Gentlemen of the Centre, that you have no deeper interest in the matter, and have no greater stimulant to act on, than we from the far East have.

We have met at a time when we have learned, in a variety of ways, what are military necessities. You, as wise men, have come here to-day to prepare for the military necessities of the future, if any such shall arise. A population on these inland seas, and along your mighty rivers, lives with no guns to protect their homes or their property. Is there not a necessity—nay, have you not a right to send up that demand to the General Government that must be answered? You know, and I know, that by our treaty stipulations we are not allowed to maintain any armament on our Western waters. You know, and I know, that there are channels of communication northward, by which naval and military armaments may be sent into the northern lakes by Great Britain, while we have no countervailing facilities. And it seems to me that he who looks at this matter as a statesman, or he who loves his country, can hardly fail to come up to the necessities of the occasion, and to say that he will have those facilities which military necessities in the future may require. More than that, when, added to a military necessity which leaves no doubt as to the wisdom of what you propose, the measure opens new channels to your commerce, and facilitates the transportation of your vast commerce to market, then the subject which has brought you together is worthy of your wisest consideration. But when you add these two together—military necessity and commercial facilities—who is there in all this broad land who will say, with the lights that we have before us, in what has already been done, that our Government will not, at the appropriate time and to the extent of its means, give a helping hand to open up the vast resources of this great country?

Gentlemen, I invite your attention to the deliberation of the questions that shall come before you. I trust, as I know, that you will bring coolness, care, and candor to those deliberations; that you will rise above all local interests, and come up to the importance of the occasion. Do that, and this shall be the starting point, and this the starting hour, in a long, and great, and glorious career of the mighty West. It will give an increased military strength. It will add new facilities to commerce, and it will bind us still closer together in harmony, in peace, in union and liberty forever.

Mr. King, from the Committee on Nominations, etc., now submitted the following

RULES OF THE CONVENTION.

1. The States, Territories, and the District of Columbia, shall be called over; and the delegations, through one of their number, shall report a written list of the delegates in attendance, with their respective residences, as far as practicable.

2. Each delegation will appoint one of its number to respond to the Chair, on casting the vote of its State, Territory, or District.

3. A vote by States may be called for by any three delegates; and in such case each State shall give the same number of votes that it is entitled to in the Electoral College, in the election of a President of the United States; and each Territory, and the District of Columbia, shall be entitled to one vote.

4. A Committee of one from each State shall be appointed by the respective delegations, to prepare and report such resolutions as they may recommend for adoption by this Convention; and all propositions and resolutions presented to this Convention shall, before being acted upon, be referred to said Committee.

5. The Convention shall be governed in its proceedings by the rules of the House of Representatives of Congress, so far as the same may be applicable.

Mr. Spaulding, of Ohio. Before the vote is taken, I would like to know if any provision has been made by the Committee in regard to the casting of the vote of a State when there is a division among the delegates?

Mr. King, of New York. There has been no such provision made; we left it to each delegation to settle that question, and to present by majority, if it can, the vote of the State.

Mr. Spaulding. I move that the report be committed to the Committee, with instructions to provide that the vote of the State shall be given in proportion to the number of members on each side of any given question.

Mr. Washburne, of Illinois. Mr. President:—I desire merely to make this suggestion touching the motion of the gentleman from Ohio: That, in order to save the time of the Convention, instead of this matter being recommitted to the Committee, he make his proposition in the shape of a motion to be decided by the Convention here.

Mr. Spaulding. I will withdraw my motion.

The report of the committee was thereupon concurred in.

The President. By the resolutions just adopted, it is provided that the delegates from each State shall appoint one person to prepare and report such resolutions as they may recommend for adoption by this Convention; and all propositions and resolutions presented to this Convention shall, before being acted upon, be referred to said Committee. The various delegations from the States will, therefore, at their earliest convenience, report to the Chair one of their number to constitute said Committee.

Mr. KING. In behalf of the Committee on Organization, I ask that they may be continued during the session; for matters may arise which will require all the attention that they will be able to bestow.

The PRESIDENT. Is that the pleasure of the Convention? It is so ordered.

Mr. BEARDSLEY, of Wisconsin. I propose that the matter be postponed till to-morrow morning. There are from the State which I represent at least one hundred delegates, and they are scattered all around this tent.

Mr. MILLER, of New York. I move to take a recess for an hour.

Mr. KELLOGG, of Illinois. It will be impossible to get the Illinois delegation together in half an hour, or three-quarters, or an hour. My friends are anxious that this matter may be postponed; and I would suggest that we have a meeting of our delegation, which is large, at eight o'clock this evening, when we propose to do business of that character. I believe it will expedite business to take abundant time to prepare such measures, and by to-morrow the delegates will be prepared to act.

Mr. SPAULDING. The delegates from distant States would like to have this matter attended to as expeditiously as possible. Most of the delegations have selected their members for this Committee, and surely the Illinois delegation ought to be prepared in as short a time as those who come from a distance. We desire to accomplish our work by to-morrow afternoon, and go home. Some of us must go home at that time. We desire that this business should be done as expeditiously as possible, so that the Committee may be enabled to report to-morrow. I merely throw out the suggestion, and hope the motion to postpone the appointment of the Committee will not prevail.

Mr. D. C. LITTLEJOHN, of New York. I rise to support the motion of the gentleman from Wisconsin. I have no doubt that many members are prepared with resolutions which can be disposed of by reference to this Committee, which will be announced in the morning. It will be impossible to convene all the delegates here within the next hour, and I hope, therefore, the motion of the delegate from Wisconsin may prevail.

The motion to postpone the announcement of the Committee on Resolutions till to-morrow morning was agreed to.

Mr. CHAMBERLAIN, of Ohio. The effect of the measure just carried will be to prevent our getting a well-digested set of resolutions. Our Committees, of course, cannot meet until to-morrow, when we have to go home. I move that the vote just taken be reconsidered.

A DELEGATE. How did you vote?

Mr. CHAMBERLAIN. Several gentlemen around me here misunderstood the question, and I ask that the vote be taken again.

Mr. ——. By the rules of the House of Representatives, which we have adopted for the government of this body so far as applicable, the motion must be made by one who voted with the majority.

Mr. ——, of New York. I will move to reconsider the vote just taken.

The motion was rejected.

Mr. ——. In order to facilitate business, I would move that the Committee on Resolutions meet at nine o'clock this evening, at the Sherman House.

The motion was agreed to.

Mr. CHAMBERLAIN. Now we stand in this light: We have to appoint a Committee to-morrow to transact certain business, to transact which it has been called together to-night. That is the action of the Convention, if I understand it. We cannot do anything until we are properly organized, and we are restricted from organization by this resolution.

Mr. VILAS, of Wisconsin. The gentleman mistakes the position of the question, I apprehend. I understand that the several State delegates make the appointment of this Committee, and that Committee will be reported to the Convention to-morrow morning. But this Committee can meet to-night at nine o'clock, because this Convention has no power to reject or confirm it. I see no difficulty then.

Mr. STRONG, of Missouri. That there may be no misunderstanding in this part of the assembly, in reference to the appointment of this Committee, I wish to call attention to the time when the several delegations should act, and when the Committee should act. It seems to me a matter perfectly plain. The delegates can come together immediately upon the adjournment of this Convention, and appoint persons to act upon this Committee; then this Committee, thus appointed, by meeting at nine o'clock to-night, can be prepared to report to the Convention in the morning. The resolution, which the gentleman on my right refers to, was the resolution not requiring the delegations to report the names of the members of this committee to-night—giving them time till to-morrow morning to do that.

Mr. Ross, of Wisconsin. I move that when this Convention adjourn, it be till nine o'clock to-morrow morning.

The motion was agreed to.

Mr. Ross. I now move to adjourn.

The motion was rejected.

Mr. Cowan, of the District of Columbia. I move that Martin Bishop, of Lafayette, Indiana, be admitted as a member of this Convention.

The motion was agreed to.

Mr. Taylor, of Minnesota. The call under which this Convention assembled, I believe, has not been read to the Convention. I would respectfully ask that the original call of the Convention be now read.

The call was thereupon read by the Secretary.

Mr. Taylor inquired if a resolution was in order?

The President said that it was in order.

Mr. Taylor. The rule, passed here to-day, requires that resolutions shall be referred to the Committee before they are acted upon. I ask whether a resolution introduced here can properly be debated without that reference first being made.

The President. I think not.

Mr. Spaulding. As resolutions are in order, I will propose a resolution as follows:

Resolved, That the subject of a ship-canal around the Falls of Niagara is one that pre-eminently demands the consideration of this Convention.

Mr. Nevins. I move that it be referred to the Committee.

Mr. Spaulding. I move that the resolution, after being read, be laid upon the table until the Committee is appointed.

The President. It is the impression of the Chair that the only action that can be taken, is to receive the resolution and lay it upon the table; and, when the Committee shall have been announced to the Convention, refer it, without debate, to that Committee.

Mr. Spaulding. I do not, at this stage of the proceedings, take any exceptions to the ruling of the Chair, but I desire to say that I will not consider myself precluded from afterwards doing so; for, in the Committee it was understood that the resolutions might be debated.

Mr. Taylor. I propose the adoption of the following resolution:

Resolved, That the construction of a Northern, a Central, and a Southern railroad from the Mississippi River to the Pacific Ocean, is properly a subject of national cognizance; but the enlargement of canals within the limits of States east of the Mississippi River is properly the subject of cognizance by the States only.

The resolution was laid upon the table.

A DELEGATE from Iowa. I move that the remainder of the afternoon be devoted to hearing addresses from distinguished persons present.

The motion was carried, whereupon

Gen. WALBRIDGE, of New York, being loudly called for, came forward and spoke as follows:

I shall not suffer even your partiality, or the kind presentation of the President, to cause me to obtrude myself upon your indulgence, for this simple reason: This Convention was drawn together for a practical purpose, and the men composing it have come thousands of miles to accomplish a specific object. I cannot, therefore, speak on any other question, except the great question which concerns us all; that is, for opening the line of communication between the East and the West, which I want you to accomplish by practical resolutions. This is not a meeting then, for any other than a practical purpose; and, while I should be pleased to listen to the addresses of other gentlemen present, and to address you myself, I do not, in accordance with my convictions of propriety, think proper to absorb your time. Nevertheless, since all practical questions, tending to the opening of communication between the East and the West are deferred till to-morrow morning, I will bring your attention to the considerations that are involved in the great struggle that is now going on, in behalf of constitutional government, within the limits of the United States; and, if any man exists, who doubts as to what will be the final termination of that struggle, let him calculate what are the elements in the respective forces that are engaged, and then he will have no difficulty in arriving at a triumphant result. It is not that we have a population of twenty millions to their six, who act with a concentrated energy worthy of a better cause, but this disparity of numbers must eventuate in giving us the ultimate triumph. But there are other things besides; they represent an article of clothing, that is, cotton, and to surrender cotton, is an inconvenience to us; we represent corn, and to surrender that, is to surender life itself. If I was simply speaking to Chicago, I should say that while corn is king, Chicago is its chosen throne.

Mr. President, and Gentlemen, I find myself beguiled beyond what I intended. If this meeting was for the practical object of constructing a line of communication by large internal canals, between the Mississippi river and the Atlantic coast for commercial purposes, the General Government has no right to do it; but if it be for military purposes, it is their imperative and bounden duty, and as such, I shall advocate this measure.

Gentlemen, there is one feature that marks this struggle in which we are engaged,—the constantly-repeated intimations of foreign powers that they will interfere before it is over,—intimations that are given, because they desire to see this government dismembered. But when you shall have opened this line of communication, by which your corn can be placed in the markets of Europe, you have a guaranty for perpetual peace. You will have accomplished more than Ministers, and Presidents, and Vice Presidents, have the power to accomplish; because, Gentlemen, population is contingent upon the means for its support, and so long as you control those elements which feed the people, you control their destinies. Therefore, when this great West, when this State of Illinois, not yet populated, and capable of augmenting her population tenfold,—when all this vast region, stretching to the Rocky Mountains, shall have become filled with an industrious population, rallying beneath the old flag that covered our Fathers of the Revolution,—when we shall have reached that period, you will supply food, not only to this continent, not only to Europe, but to that far distant land beyond the Ganges, which opens that great Oriental trade which has enriched every people who have engaged in it. It is to be brought across this continent by means which you have this day thought fit to originate.

Mr. Drake, of St. Louis, was next called for.

The President. I take pleasure in presenting to you a friend from that honorable State of the West, that bright star which shall shine brighter and brighter unto the perfect day. Mr. Drake, of St. Louis.

Mr. Drake spoke as follows:

Mr. President: On behalf of that noble city which I have the honor here to represent, I thank you, Sir, for the testimony which you have borne to her noble and genuine loyalty. One myself, of a population of probably one hundred and three-score thousand, I feel, Sir, that I can give credit to St. Louis for her loyalty, without assuming to myself any vanity. When you reflect on all the circumstances which have surrounded St. Louis, from the day that the traitor Governor of Missouri was inaugurated in January, 1861, to the present time,—if you could only, my friends, know them as we who have lived there have known them,—you would say with me that no city in this land deserves higher praise for the loyalty of her people than St. Louis. But, Sir, St. Louis, in my person at this moment, in her delegation which occupy seats in your midst, greets you patriots of the land in this assembly, convened under circumstances such as have probably not been known in the world's history before. With you, and with all, to whom that venerable flag is dear throughout this land, St. Louis extends the arms of greeting and affection, desiring to live with you, and to die with you, under its resplendent folds. Under circumstances which have brought this body together, on the margin of this fair lake, in this queen city of the valley of the St. Lawrence, one part of this nation is covered with blood, in a conflict for the nation's life. Another part is assembled here, by its representative men, to consider measures which shall make for the present an enduring prosperity, in peace, or in war, a mighty advantage. Take any other nation upon earth, and no such thing is possible; no other people can fight the battle we fight, with one hand; and, with the other, scatter over the earth the blessings of life. If I had never been proud of America before, I should be proud of her now, and so should you. We have a right to be proud of that, which ennobles and glorifies us as a people. We have a right to be proud of this glorious Valley, which, from these points below, sheds its waters over Niagara; and of that other glorious Valley, which, rich and fertile in the North, casts its treasures of living waters beneath the sun of the tropics. We have a right to be proud of all that dignifies and elevates us as a people; gives us power at home, and consideration abroad; and amid all the developments that are calculated to produce these results, nothing, probably, in our history, has yet exceeded the developments of this day.

The speaker then proceeded to speak on the topics of the day; and, as his remarks did not immediately bear upon the objects embraced in the call of the Convention, they are omitted.

At the conclusion of his remarks, on motion of Mr. Ruggles, of New York, the Convention adjourned until ten o'clock A. M. of the following day.

Wednesday, June 3, 1863.

At ten o'clock, the President took the Chair, and announced: The hour to which this Convention adjourned having arrived, you will come to order. I have the pleasure to introduce the Rev. Dr. Clarkson, of Chicago, who will open the proceedings with prayer.

PRAYER.

We praise thee, O God; We acknowledge thee to be the Lord. All the earth doth worship thee, the Father Almighty. Heaven and Earth are full of the Majesty of thy Glory. Day by day we magnify thee; and we worship thy name ever, world without end.

O Lord God of Hosts; Let thy blessing rest upon this people who are assembled here before thee. Direct them this day, and always, in all their deliberations, with thy most gracious favor, and further them with thy continual aid; that in all things, begun and ended in thee, they may glorify thy Holy name. Prosper their consultations to the advancement of thy honor; and to the safety and welfare of this great nation; so that truth and peace, religion and piety, may be established among us for all generations.

Oh Lord God Almighty, of thine infinite power we beseech thee to heal the desolations, and assuage the calamities of war; and restore, we pray thee, the broken unity of this suffering people, and hasten the coming of that blessed time, when liberty and righteousness shall reign supreme on all the earth. All which we ask in the name and for the sake of Christ Jesus, our Lord and Saviour. Amen.

The PRESIDENT. The SECRETARY will read the list of the Committee on Resolutions; which was as follows:

Missouri,	Mr. Henry Hitchcock.
Kansas,	" J. C. Trask.
Wisconsin,	" Orsemus Cole.
Maine,	" John Lynch.
New York,	" Samuel B. Ruggles.
District of Columbia,	" Peter Parker.
Michigan,	" Alexander Lewis.
Minnesota,	" J. W. Taylor.
Iowa,	" Richard B. Hill.
Illinois,	" William Kellogg.
Ohio,	" A. G. Riddle.
Connecticut,	" Samuel Gold.
Massachusetts,	" William Hilton.
New Hampshire,	" Thos. M. Edwards.
New Jersey,	" Joseph C. Jackson.
Indiana,	" G. W. Julian.
Rhode Island,	" R. J. Arnold.

Mr. SPAULDING, of Ohio, moved to take from the table the resolution which he offered yesterday, for the purpose of offering a substitute.

The motion was agreed to.

Mr. SPAULDING then moved as a substitute, the following preamble and resolutions, which were referred to the Committee on Resolutions:

WHEREAS; It has been ascertained by a careful survey and estimate, made by Capt. W. G. Williams, of the United States army, under the direction of the Topographical Bureau of the General Government, that a ship-canal on the American side, between Lakes Erie and Ontario, to pass vessels of from 1,200 to 1,500 tons burthen, can be constructed on a line of from seven to fourteen miles in length, and at a cost of from two and a half, to five millions of dollars, depending upon the line adopted; and

WHEREAS; The construction of such a canal is demanded by the entire Northern frontier, as a military defense; admitting, as it will, the passage of gunboats of a capacity equal to any that can be put upon Lake Ontario by the British Government, through her already constructed St. Lawrence canal; and

WHEREAS; Through such a canal, admitting, as it will, the free lake-navigation of steamers and vessels of a large class, the products of the Great West can, cheaper than by any other channel, be poured out on the Atlantic coast for distribution to the markets of the world; therefore,

Resolved, That the Committee recommend, in connection with other measures, the immediate construction of a ship-canal around the Falls of Niagara, as a Government work, and at the national expense. (Appendix, C.)

Mr. FOSTER, of Illinois, submitted, on behalf of the Board of Trade, the Mercantile Association, and the citizens of Chicago, a report as to the necessity of a ship-canal between the Mississippi river and the Atlantic, and proposing the following measures:

First; To improve, under the authority of the General Government, by slack-water navigation, the Illinois and Des-Plaines rivers, by constructing a series of locks and dams, 75 feet in breadth by 350 in length, and 7 feet in depth; and to enlarge the present Illinois and Michigan canal to a like capacity; so as to admit the passage of gunboats, and the largest class of steamers from the Mississippi to the Lakes, and *vice versa.*

Second; To enlarge, under the same authority, the locks of the Erie and Oswego canals of New York, to such dimensions as to pass iron-clad gunboats 25 feet wide and 200 feet long, and drawing not less than 6 feet 6 inches of water; by which twin improvements gunboats may be passed, by an interior route, from New Orleans to Chicago, Buffalo, Oswego, New York, Norfolk, Richmond and Beaufort, a distance of 4,300 miles; besides placing under the control of the naval power of the Government, the whole navigable system of the Lakes, and the Mississippi.

As auxiliary to, and as forming a part of the report, Mr. FOSTER also submitted a special report of the survey of the Illinois and Des-Plaines rivers, executed by Messrs. Gooding and Preston, engineers of the highest capacity; and asked that it be referred to a Special Committee of five, to be appointed by the Chair, and to be composed of engineers eminent in their profession, who are to be required to report to the Convention as to the practicability of the work, and the correctness of the estimates. (Appendix, A. and B.)

The reports and estimates were, under the rule, referred to the Committee on Resolutions.

Mr. RUGGLES, of New York, reported from the Committee on Resolutions, the following resolution submitted by the delegates from the District of Columbia, with a recommendation that it be adopted:

Resolved, That no subject or resolve not clearly germain to the objects for which this Convention has been called, shall be admitted or discussed during the sessions of the same.

Mr. LITTLEJOHN, of New York, inquired whether it was the object of the Convention, in reporting this resolution, to declare discussion in reference to the Niagara ship-canal out of order?

Mr. RUGGLES, of New York, replied in the negative, stating that such an idea was not thought of.

Mr. LITTLEJOHN wished it to be fully understood by the Convention, that no such technical construction was to be given to the resolution.

Mr. RUGGLES assured his colleague that the Committee stood on no technical points. The Committee had no wish to exclude discussion on the subject of the Niagara ship-canal.

Mr. LITTLEJOHN explained his meaning. The call of the Convention referred to the enlargement of canals. The Niagara ship-canal project involved the construction of a new canal; and was not, therefore, an enlargement—so that a technical objection might, under the resolution, exclude that subject. Under the explanation of the Chairman of the Committee, he was quite content.

Mr. RUGGLES stated that the resolution had been brought forward by the delegation from the District of Columbia, and its principal object was to exclude political discussions, which were at this time thought unnecessary.

The question was taken, and the resolution was adopted.

Mr. RUGGLES, of New York, from the Committee on Resolutions, reported back the following resolution, offered by Mr. Taylor, of Minnesota:

Resolved, That the construction of a Northern, a Central, and a Southern railroad, from the Mississippi river to the Pacific Ocean, is properly a subject of national cognizance; but the enlargement of canals, within the limits of States, between the Mississippi river and the Atlantic ocean, is properly a subject of State cognizance alone.

He stated that the resolution had been considered by the Committee, and that he was instructed to make the following report:

"The Committee on Resolutions, to whom was submitted the subjoined resolution, report that, in their opinion, it ought not to be passed by the Convention."

Mr. WASHBURNE, of Illinois, moved to lay the report and resolution on the table.

Mr. TAYLOR, of Minnesota, asked the delegate from Illinois to withdraw the motion for a moment, that he might make an explanation.

Mr. WASHBURNE, of Illinois, withdrew the motion.

Mr. TAYLOR then spoke as follows:

The most potential argument, Mr. President, in favor of the resolution, which has not been so fortunate as to receive the sanction of your Committee, is visible in the spot, and in the scenes, and the associations where we stand, and by which we are surrounded. What is Chicago at this moment?

What is the great physical, commercial, and cosmopolitan problem here presented for our consideration? It is the marriage of land and water. It is the marriage of the iron-rail to the great water-communication of this Mediterranean of the continent. Chicago stands to-day the apparent queen of all these Great Lakes; because here, within the sound of this Convention, converge twelve lines of railway, the highest triumph of modern civilization. And it is the transhipment of the products of the people westward of the Lakes, penetrated, and reached, and developed, by these twelve lines of railway, that has given to these Great Lakes a commerce which is at this moment greater than all the external commerce of the United States; and which has made the question of ship-canals, and of every possible auxiliary by canals, to this great water-channel, the vital question of the hour.

And now, Sir, in behalf of a western State, of a frontier State,—of a State reeking with Indian massacres, with the scenes which sixty years ago only, were enacted on the very spot where we stand to-day,—I simply ask the citizens of those States to consider well the questions, whether the enlargement of the canals eastward to the sea-board; whether the construction of a ship-canal around the Falls of Niagara; whether any and all means further to organize and develop this great internal commerce, will not be better advanced, will not be more surely secured, by planting ourselves, as a great nation, on the idea of Pacific extension, of railway extension westward from the margin of the Mississippi river, on the broadest national scale, to the Pacific coast; and whether by keeping singly before us, and presenting singly to the American people, that great, broad, fundamental idea, you will not do far more to advance these great interests, than by frittering away your strength upon an indefinite series of canal-projects within States, which are amply competent to establish and effect them for themselves.

I may be reminded here, as I have been reminded in the Committee, that the Pacific Railroad question is settled by the action of a dismembered Congress. I deny it. I deny that the construction of a railway from the city of St. Louis, whatever may be its collateral termini, to the city of San Francisco, is the solution of this great national question. By no means, Sir. It is but a partial solution. It is but the beginning of the end. It is but the entering wedge of the great question. It is the anticipation and determination of the American people, that it shall be settled, and re-established, on a foundation as broad as our glorious nationality. I look to that great river of the West beyond you. I see a railway projected from the mouths of the Ohio and Missouri rivers to the bay of San Francisco. I float down that mighty Amazon of the North, past the frowning batteries of Vicksburg, over which our glorious flag is soon to wave in triumph; past every mere temporary demonstration of treason against this glorious Republic, to that mighty city at the mouth of that river,—to that great crescent metropolis of the Gulf,—to that city over which once the iron will of a Jackson predominated for the benefit of the country, against an invading enemy, and where the iron will of a Massachusetts Butler again restored peace. I say in behalf of the North;—I say in behalf of the bleeding South,—of the trampled, outraged Union sentiment of the South,—that when this Government is restored, when the flag of the Union once more asserts its supremacy over the length and breadth of the land,—Chicago will be the first, Illinois will be the first, all the North-West and all the North-East—New England joining nands with us—to say, that what the Legislature of 1862 enacted, on behalf of the great central line of Middle States, the measure of a railway to the Pacific Ocean, from St. Louis to San Francisco, —is not and shall not be the ultimatum of that question; but, when the erring and outraged sons of the South return to their allegiance, and again join hands with us, to support forever this glorious Commonwealth; then, a Southern Pacific railroad, from New Orleans to San Diego, will be a measure of peace, of conciliation, and of national unity, which the North-East and the North-West will cordially and fraternally join hands in establishing.

For, I do not for one share the sentiment, that we are to take advantage, for any mere local or material purpose, for any mere question of canals or internal improvements, of the enforced absence from the halls of Congress at Washington of the true men of the South, who are now the victims of the worst despotism that ever defaced God's footstool. I want the Southern States represented there. I want, when the power of the Government is again asserted as, in God's good time and pleasure, it will be, the men of the South, like Andrew Johnson, of Tennessee, to have seats on the floor of Congress. I want the brave, true-hearted men of East Tennessee, the true, loyal element of Northern Alabama, of Northern Georgia, of Western North Carolina, of Louisiana, and of Texas, to be represented there.

Feeling confidence, as I do, in the progress of the West, in its power present and power to come, I am in no eager haste to force conclusions on these questions. I am perfectly willing to bide the time of the restoration of the Government, when we shall again meet in fraternal council to settle these great questions of internal improvement.

Mr. Taylor proceeded, at considerable length to elucidate his views on the Pacific railroad question, and on the canal question, contending that while the former was a proper subject for national cognizance, the latter was a proper subject for State cognizance only; that Illinois, with her great wealth and resources, should not stand, like a pitiful mendicant, asking Congress to do what she should do for herself; and that New York should not be here in support of that contemptible effort.

Mr. Washburne, of Illinois, responded.

Mr. President: As I moved that the resolution of the last speaker should be laid upon the table; I desire, now that the gentleman has been heard, to say one word as to the reasons which induced me to make that motion. This Convention has met here in the midst of one of the most stupendinis wars of which history makes mention, at a time when the nation is struggling for its existence; when the hearts of the people are wrung with anxiety and anguish; when all the loyal States have sent here their most eminent and distinguished representatives to consult and deliberate on the common defense and general welfare; and I thought it out of place that a resolution should be introduced here, on the very threshold of the proceedings, which would nullify everything that we might undertake.

I therefore made what I conceived to be the appropriate motion, —to lay the resolution summarily on the table. I withdrew that motion out of courtesy to the mover of the resolution, that he might be heard in its defense. He has been heard, and the Convention can determine on the weight of his arguments. What are they? He travels off on the question of building a Pacific railroad, apparently forgetting that the same power which we shall ask to open up this water-communication from the Mississippi to the Atlantic, has already made the most ample and stupendous grants to accomplish the very object, which the gentleman seems to have so much at heart. I do not believe, Mr. President, that we travel that way at this time. The people of Minnesota have the same interest, which the people of the entire North-West have, in the great measure which this Convention proposes to consider. That gem of a State of the North-West, with its genial skies and fertile soil, with its enterprising and industrious population, is identified with her sister States of the North-West. I do not believe her farmers now purpose finding a market for their wheat over the gentleman's railroad, at Puget's sound, instead of the Atlantic coast.

What is it that the people of the five wheat-growing States of the North-West want? How can the present prices of freight be reduced, unless we enlarge the channels of communication? That must be done; and then Minnesota, like all the North-West, will have the benefit of it.

I do not depreciate the importance of a railroad across the Rocky mountains to the shores of the Pacific; but certainly that cannot take away from our project its vast importance. The gentleman has spoken in regard to measures, which he says ought not to be considered by the Congress of the United States as it now exists. He has spoken of the enforced absence of men from the halls of Congress; and has argued that, therefore, the people of the loyal States, through their representatives, should not enact such legislation as they may see fit, because traitors have left their seats. As the humblest member of that body, I tell the gentleman from Minnesota, that my action will never be governed by any such considerations. If they are not there, let them take the consequences. Those of us who are there will legislate for the benefit of those whom we represent, and of the whole country. He has said that the States of New York and Illinois should not present themselves as pitiful supplicants for favors at the hands of Congress. Does the gentleman understand where Illinois stands today, with her 150,000 brave men making the earth to tremble beneath their tramp? With her troops, carrying victory inscribed on all their banners, so recently triumphant on so many bloodstained fields, led by their brave and unconquerable Illinois General,—Grant? Illinois a mendicant! Never, never. Does not the gentleman know that there are great national interests, over and above all local interests, involved in the propositions which this Convention is to consider? This Illinois and Michigan canal is to be enlarged, not only to give an additional outlet to the products of the industry of our vast West; but, as a more ready means of defense of the immense commerce that floats on our great inland seas; and to the defense of the towns and cities on their shores. The national interest requires that you should have this canal so enlarged as to permit the passage of gunboats from the Mississippi to the Lakes, to be ready for any emergency, and for all comers.

Is not the Government interested in that? Are not the gentleman's constituents in Minnesota interested in that? Why should not we of the West have some attention paid to us by Congress, for our defense in time of war? Your President, yesterday, in his admirable and beautiful address, told you that, under existing treaties, we have no armament on these Lakes—substantially none—only one armed vessel; while Great Britain, in the shortest period of time, can make the whole Lakes bristle with her armed vessels. That is one of the reasons why the Government should take hold, and open up the channels of communication; so that you can get your gunboats to the Lakes if necessary; so that you can send a gunboat from Albany, through the Erie canal to Lake Erie;—so as to be prepared to meet any emergency that may arise. Let me tell you here, with all earnestness and seriousness, that, unless the arm of that haughty power—Great Britain—shall soon be stayed in her aggressions on our commerce, the clash of arms will be heard in another direction, from where it now resounds. And when it does come, are we here to be defenseless? Is this Queen City of the Lakes; are all the towns and cities which dent their shores, to be left powerless and defenseless? I say, no. During the short time that I have been in Congress—as the President can bear witness—my votes have always been given for the largest appropriations for defenses on our Eastern coast; and is it to be wondered at, that I demand that what we have so freely given to the East shall be now as freely given to us?

These are the military considerations. Then look at the incidental commercial considerations. Let me ask the farmers of Illinois, of Iowa, of Indiana, of Ohio,—the mother of us all—of Minnesota, and of Wisconsin, how the question stands. It may be easy for the gentleman from Minnesota (Mr. Taylor) to talk as he does; but, if he raised corn at ten cents a bushel, and it cost twenty-five cents to get it to market, I think he would look on the other side of his ledger. He has spoken of Chicago, and the interests which this city has in the matter. Why, how many, in this vast throng which I address, are interested in the city of Chicago? Sir, it is not the interest of Chicago that is represented here; it is the interest of the whole country. It is the interest of labor. And I am gratified to see so many of the farmers here to-day, for they are the men who are interested in the matter more deeply than any other.

But, Mr. President, I will not detain the Convention any longer. I desired merely to state briefly the reasons which had induced me, at the very threshold, to move to lay the gentleman's resolution on the table. We have not met here as a Railroad Convention, or as a Convention to take into consideration any other subject but that which is mentioned in the call. However we may agree with the first part of the gentleman's resolution, in regard to the Pacific railroad, and the duty of the Government in that respect, we certainly cannot permit the resolution to be considered. I therefore renew the motion to lay it on the table.

The question was put, and the report and the resolution were laid upon the table.

The Chair. The Secretary will announce the arrival of a delegation of true and loyal men from the State of Kentucky, as members of this Convention, together with the persons they have delegated to act on committees.

The Secretary read the names of the Kentucky delegation, as follows:

Thomas S. Page.	Dr. H. Rodman.
M. G. Knight.	Wm. Terry.
W. J. Morton.	R. H. Cramp.

Committee on Resolutions—Dr. W. J. Morton.

For Vice President—Wm. Terry.

The Chair requested the gentlemen named to take seats upon the platform.

Mr. Spaulding, from the Committee on Organization, said that the Committee had consulted upon the propriety of restricting speeches to a given time, but had come to no decision. He would, however, assume the responsibility of moving that delegates speaking be restricted to ten minutes, the rule to apply for this day.

Mr. Spaulding's motion was then put upon its passage, and prevailed.

Mr. WASHBURNE, of Illinois. I have the pleasure of announcing the arrival of a delegate from Massachusetts, Hon. H. L. Dawes, a member of the last and present Congress, and one of the most earnest and able advocates of the Canal measure.

Mr. RUGGLES, of New York, offered the following resolutions:

Resolved, That in view of the immense increase in the products of the States adjacent to the Lakes and the Upper Mississippi, and the necessity of providing for the defense of the national commerce on these waters, this Convention hold it to be the imperative duty of the General Government to provide any necessary facilities for cheapening and protecting said commerce, by adequately enlarging, and improving the canals and rivers, now connecting the Lakes with the Hudson and with the Mississippi rivers.

Resolved, That the proposed advantages, military and commercial, of a canal around the Falls of Niagara, be also commended to the careful examination of Congress.

Mr. RIDDLE, of Ohio, offered the following resolution as a substitute of the second:

Resolved, As the sense of the Ohio delegation, that our member of the Committee on Resolutions be instructed to keep the subject of a ship-canal around the Falls of Niagara prominently in view, in connection with the system of improvements to be recommended by this Convention.

The PRESIDENT requested the Secretary to read the original resolutions, and the substitute for the second, which was done.

Mr. A. G. RIDDLE. It will be discovered that there are two propositions reported by the Committee, but both are not upon the same basis. That in regard to the ship-canal around the Falls of Niagara is placed in a subordinate position. I should desire, only so far as any personal views which I entertain as an individual are concerned, that it be placed upon precisely the same basis that the first general propositions are. An intelligent stranger, who would first unroll the map of North America, would be struck, perhaps, with nothing more directly than with the immense system of great inland waters. Following the map he would necessarily see whether there was a navigable outlet existing from the Great Lakes to the Ocean, and he would be struck by the entire obstruction of navigation by the Falls of Niagara, between Lakes Erie and Ontario. The first proposition contemplated that a great population should spring up upon the shores of these inland seas; and their configuration, their location, and their relation to the rest of the continent, seem to have marked them as the destined abode of a great people; it was in fact the promise of their creation, that if they were not the greatest, they should, at least, form one great element of the nation. The stranger would ask whether it was possible to secure navigable facilities from the Great Lakes to the Ocean. He would try to ascertain what would be the degree of practicability of connecting this great chain of lakes, long before he dreamed of the union of the waters of the Mississippi with those of the Upper Lakes, or even the construction of the Erie canal.

He would determine that it was the duty of the sovereign to complete what nature had thus left incomplete. If he looked further, he would find that the sovereign of the soil on the northern side had acted on that natural hint, and had completed what nature, for some purpose, had left incomplete, and had already constructed a ship-canal. He would be utterly amazed that a great nation like our own, which had planted its footsteps on the further shores of those Great Lakes, was still content to grope through that canal, which the British Government had constructed on the other side, and to pay tribute therefor. He would be utterly amazed that the great nation, on the south side of these now unnavigable waters, had not acted on the teachings of nature herself, and followed the example of her great rival by constructing for herself, on her own soil, a ship-canal around the Falls of Niagara.

A population has sprung up upon the shores of these Lakes, greater than the entire population of the United States at the commencement of the war of 1812. There are cities strewed all along those Lakes, larger, some of them, more populous, and more powerful, than any city upon our western continent at that time.

May I not say that this is the first great link in the plan of lake-navigation? However great, or however important the other propositions are, I submit that, if not entitled to the very first place in the proceedings of this Convention, it is entitled to be placed upon the same basis as the other. This is the proposition you have to pass upon.

I would like to have said something upon the main proposition; but I do not propose to do so now, as the question is not before the Convention. My voice has been raised in another place, and my vote has been given, for the general proposition that underlies the first branch of the report.

Mr. SPAULDING, of Ohio, followed. It was distinctly understood by this Convention, that I am peculiarly in favor of that great work of internal improvement—a navigable ship-canal around the Falls of Niagara; and I now avow, for the first time, that I am equally in favor of a ship-canal from the Gulf of Mexico to the Gulf of St. Lawrence. How is this to be accomplished, except by a canal around the Cataract of Niagara? Is there anything lying under the surface which we do not now understand? If so, this Convention is acting in the dark. I am a practical man. I have but a few words and those short ones. It is said that the great work is a military necessity. So it is. This Convention itself is a great military necessity; and its assembling at such a time as this, for the purpose of promoting the common defense for all time to come, will produce a more stunning effect on the rebellion than several victories by our armies. I would desire that those who are in favor of the enlargement of the New York canal, and those who are in favor of the enlargement of the Illinois canal, should meet each other half way. The whole lake-shore interest have much at heart in this particular work—the construction of a ship-canal round the Falls of Niagara. I do not desire to speak in words of menace; I only desire to speak in words of harmony. This measure requires all the aid we can give to it. Now, do not let us, by our course to-day, cast from us one single vote in Congress. We can be harmonious, and give our energies to the enlargement of both canals, and the construction of a ship-canal, thus making a practical ship-communication from Gulf to Gulf; and then, with the blessing and the smiles of God upon us, we must be successful.

Mr. Ruggles. The Committee, Mr. President, is attacked by the gentleman who has just sat down, and this Convention is told that there is something lying under these resolutions, which he does not understand. I will tell him what lies under the resolutions. It is the welfare of the United States which lies under them, and the safety of its commerce, and the development of its resources, and the guarding of this commerce, and these developments from foreign diversion; nothing else. Who ever heard before of the Gulf of St. Lawrence in this connection? Has not our whole struggle been to prevent this Canadian diversion? A principal aim of the memorials to Congress, and the documents issued by the Boards of Trade, was to prevent our internal commerce from being diverted from our own water-courses and sea-ports, into those of foreign nations. I do not say that this result would certainly flow from the construction of this ship-canal; but, to say the least, it is possible. But, I do not mean to take ground against the ship-canal here, and perhaps not elsewhere; we cannot, however, close our eyes upon the magnitude of the consequences involved. The question, whether the hundred millions bushels of grain,—soon to be multiplied to a thousand millions—is to be poured into foreign ports, rather than into our own, is a national question of the gravest import, and one which deserves the consideration, not only of this Convention, but of Congress. Hence, the form of the second resolution, which asserts that the military and commercial advantages of the measure deserved the attention of the General Government. I admit that such a work might produce some saving in transportation, and might add to the military security of the country; but, at the same time, we should keep our eyes fixed on the preservation of our commerce.

Mr. Henry, of Iowa. I offer an amendment to the resolution of the gentleman from New York, which I suppose he will have no objection to accept. I offer to insert after the word "Niagara," the words, "and, also, so to improve the Rock Island and Des-Moines Rapids of the Mississippi river, as to pass gunboats over them." This is part and parcel of the great chain of national work. The Mississippi requires attention to her defenses, and the passage of gunboats, equally with the Lakes, and I, therefore, hope the gentleman will not object to my amendment. (Appendix, D.)

Mr. A. G. Riddle. Is not that intended as an amendment to the original proposition?

Mr. J. B. Grinnell, of Iowa. I observed the smiles that passed over the countenances of the gentlemen from New York, because the word Des-Moines was mentioned. I wish the gentlemen from New York to understand that the present proposition has no relation whatever to their Des-Moines improvements. A company from that State placed some fifty locks there for the purpose of damming (though they did not mean it for *damning*, but for improving) the water-communication.

Now I wish to state, in a few words, what is the position of the State of Iowa.

Mr. President, I wish you to understand that, from the State of Minnesota, for 500 miles to the State of Wisconsin, and the State of Illinois, the productions of the soil all go down the Mississippi river; and in doing so, have to be laden and unladen two or three times. Now, suppose the project of this Convention is carried out, to construct a ship-canal in Illinois. How can you expect a vote from the State of Iowa, and how can you expect me to be sustained at home, if I vote to appropriate $100,000,000 for these purposes, unless we can have a moiety for making the Mississippi navigable, so that our gun-boats and our corn-boats can pass to and fro upon it. Now, Sir, this is our proposition: We expect to support the resolution in favor of this ship-canal, and we shall couple with it a demand that the Mississippi shall be cleared. By that, we mean that the Upper and Lower rapids may be made passable for gun-boats and steamboats. No one believes that it will cost more than a million dollars; and, if it succeeds, we shall bring to you millions of value in wheat.

I beg the Convention to bear in mind that Iowa has nearly a million of people, and fifty thousand brave men in the field. Thirty of her noble regiments are to-day before Vicksburg, if not in it, as I trust to God they are. And this is the only thing we ask; that you will believe us as in earnest, when we say, that we want an outlet to Chicago, to Buffalo, to New York, and to Liverpool, and that you will help us in making this river passable.

Why, Sir, only one-ninth of our territory is under cultivation. You gentlemen here, speculators, own it. I invite you to go over and look at your broad acres. Not one acre in ten of our country is improved, and yet we have nearly a million of people. We are not boasting; but we have less waste land in our State than any other State of the Union. It is all entered, except one or two million acres. Come, vote with us, act with us, do not sneer at us, indorse us. Incorporate this amendment with that resolution, and I believe you can then be assured of the support of the people of Iowa.

I have looked to this Convention as a means of binding the East and the West together. I predict that, in ten years from this time, the State of Iowa will send away a larger number of bushels of wheat and corn, more value in pork, and beef, and wool (and I am a wool-grower myself), than all that now passes the gates of Chicago. I say again, that I look to this Convention as a means of cementing the East and the West.

It is my custom to make an annual pilgrimage to the East, where are the graves of my fathers, from which I return more deeply impressed with the worth of the Union, and with the future greatness and glory of our country. In the heat of my passion, I have thought I could shed my blood to keep our States together. To this Convention I have looked for practical deeds; and have hoped we might love more as brothers by our meeting, and embrace like Joseph and Benjamin, though one was reared in Egypt, and the other in Canaan. Commerce, interest, and duty, may bind us with a golden chain. Our hearts are in the work, which will give value to our produce, unity to our people, and bring the Mississippi, cleared of its obstructions, in connection with Lake Michigan and the Atlantic ocean.

Mr. Leighton, of Iowa. I believe, Sir, that light ought to be shed on all questions brought before this assembly. I believe, Sir, that many of those gentlemen who smiled at the mention of the Des-Moines rapids, would cease to smile if they reflected that the Des-Moines rapids are the rapids of the "Father of Waters;" that small stream, Sir, that rises about five or six hundred miles above where I live (Keokuk), and debouches in the Mexican Gulf, fifteen hundred miles south of me; that is the stream, Sir, which the gentleman's amendment proposes to have cleared for navigation in the Upper and Lower rapids. The distance is about twenty-two miles altogether, and the improvements can be made at a cost of $1,000,000, as I am informed by eminent engineers; amongst others by Major General S. R. Curtis, who thinks that, as far as Iowa is concerned, we have no interest here at all, unless the rapids of the Mississippi are cleared out in some way or other; either by blasting out the rocks, or building a canal around them. Even if you obtain the improvements now contemplated, the larger portion of Iowa has still the impediments of the Mississippi navigation to contend with, and she cannot send a bushel of grain, or a single soldier, except over these impediments.

We think this is perfectly germain to the objects of this Convention. I know that the gentlemen from St. Louis will not laugh at us for this amendment. I know that the business of Iowa is worth, to them, some thousands of dollars per year; but I look forward to the day when, with the gentlemen from Minnesota, we will again trade with our misguided friends in the South. We want again to send our produce to the Southern planters of the Gulf States, as we did in olden times. We want the rapids cleared out, that we may extend our hearts and our right hand to the East, and our left hand and what remains of our hearts, to the South. I speak as a lover of the Union, one who would give his life, his fortune, and his sacred honor, in favor of supporting the Union, inseparably one, now and forever.

I ask gentlemen not to indulge in smiles at the mere mention of a name, as if it were that of some unknown and insignificant stream. It is the great "Father of Waters," the Amazon of the North, the main artery of this half of the continent.

We ask you to make the improvements which a young and gallant State has a right to ask; a right to ask that these impediments shall be removed, and that we may send our grain to the North and South, as well as to the East and West. Now, the transportation to New York is three times the cost of raising grain. It is even cheaper for us to burn corn for fuel, than to go into the woods and cut it, paying a man seventy-five cents per cord for his labor. We ask that our corn may be sent to the suffering millions of Europe. The railroads cannot do it. They are already overcrowded with business, and their resources more than taxed. I see upon the platform a most worthy and excellent railroad president, and he knows the truth of my statement. The railroads are waxing fat, and we of the West want a finger in the pie. I most cordially second the amendment for the improvement of the rapids.

Mr. Hubbell, of Wisconsin. I am loth to utter a word coming in the way of this Convention. But I ask the movers of these resolutions, and the gentlemen discussing them so eloquently, what is the policy in this effort to induce the Government to build canals or remove river obstructions? Is it a military necessity? We have heard it discussed as if it were only a commercial necessity. In that light I concur in all that has been said, or can be said. Is this canal through Illinois sought for on the grounds that the necessities of the Government now require this expenditure of money. We are told the nation is engaged in a gigantic war, and staggering under the weight of the blows heaped upon it. It stands, sword in one hand and pistol in the other, summoning every man in the loyal States to protect its life. Does this Convention come here now to ask permission to put its hands in the pockets of the Government to build canals? Is it essential while this war is being carried on, that this money shall be taken? If so, take it as freely as you take the heart's blood of the soldiers of Illinois and Wisconsin. If not so, spare that money. I openly ask members of Congress, if this construction of canals be not a military necessity, will they dare to put their hands in the public purse to take money and build them? I would rather burn the houses of the Sanitary Commission, or rob an hospital, for that would injure only a few sick soldiers. The abstraction of this money would wrong and injure the whole country.

For what purpose are these canals a military necessity? How long will it take to build them? Three, four, or five years? What military necessity will there be at the end of that time? This rebellion? God forbid that it should last that long. A war with England? When did we have a war with England? Fifty years ago. When shall we have another? Not for fifty years to come, except by our own volition. If John Bull had chosen to fight with Brother Jonathan, he would have done it last year, when the South had us by the throat. Now, when we have got the South by the throat, and John Bull knows it, is he going to venture his commerce against our armaments? No! if John Bull fights with us, it will be during the rebellion.

Is it a commercial necessity? I admit it, but not a vital one. The railroads for years past have transported the grain of Wisconsin and Minnesota at twelve cents per bushel. The canals and improvements of the rapids will never transport the wheat of Minnesota and Iowa as cheaply as the railroads. Can you send wheat down the Mississippi and up the canal, which can be open only half the year, and compete with the railroads, which run all the year?

Mr. Levi Blossom, of Wisconsin, suggested a point of order, that when an amendment is pending, speakers should confine themselves to the amendment, and not talk at large upon the main proposition.

Mr. Levi Blossom, of Wisconsin. I offer an amendment to the amendment now under consideration, by adding these words: "And the widening and deepening of the channel through the St. Clair Flats, and the enlargement and deepening of the channel of the Fox and Wisconsin Rivers from Lake Michigan to the Mississippi." (Appendix, E.)

Gentlemen of the Convention: It is not my purpose or intention to entertain you with a speech upon this amendment, but as it has been decided that this is a Convention for canal and river-enlargement purposes, and not a railroad convention, I propose to bring before you the proposed channel through Wisconsin, and the improvement of the channel through the Flats. Without this latter highly important improvement, we have no assurance that we can get to New York at all, or to the Gulf of St. Lawrence. Let all of us understand that it is *impossible*, and I need not further enlarge upon it.

Upon the deepening of the Fox and Wisconsin rivers, I know I shall be met with sneers. There are two natural channels from Green Bay to the Mississippi, of 250 to 300 miles in length, through a rich and fertile country, second to none in importance. Now between these two natural channels, throughout that extent, an artificial channel of one and a quarter miles mingles the waters of the Mississippi and the St. Lawrence. The like does not exist in any other place on this continent. Little is said, little is known about it. And yet they have so far succeeded now, in deepening and widening this canal, that barges carrying one hundred tons can go through. There is a natural channel running through a country some 250 miles in extent, that can be made deep enough, and wide enough, to carry gun-boats through from Lake Michigan to the Mississippi, and then by the improvement of the Des-

Moines rapids to let them down the Mississippi. It is all true. I simply call your attention to the importance of this channel, and the cheapness with which it can be built, in comparison with any other. Now, I am going to state a fact, and that is, that $4,000,000 will carry all the gun-boats of the United States from one river to the other; for we have got a channel there, made by the God of nature, which, for all practical purposes, can be made just as available for the produce of Minnesota, as for that of Iowa and Wisconsin, which never would be brought down to the mouth of the Illinois river, and thence up to Chicago. The railroads will bring that. You could by this route bring the produce of Wisconsin, Northern Iowa, Minnesota, and that of a large number of States, yet to be developed out of the district west of them.

Mr. A. G. Riddle. Are they not a portion of the rivers named in the first branch of these resolutions, as rivers connecting the Mississippi river with the Upper Lakes?

Mr. Levi Blossom. I admit they may come within the limits of that resolution.

Mr. Leighton. Some of my friends say that I have stated that "if these rapids were not improved, Iowa had no business in this Convention." I meant that commercially speaking, Iowa had no interests here; but Iowa, for any project of a military nature, has an interest here equal to that of any State of the Union; therefore, Mr. President, I wish that portion of my speech to be corrected. We ask what we think we are entitled to in a commercial sense, but when you ask if Iowa is sound for the Union, and in favor of a canal for the transport of her troops and gun-boats, you have every man, woman and child with you.

Mr. Blossom. I understand that all these resolutions are to be communicated to Congress. Now, it is with that view, and that all these people from the East and from the West may understand when you speak about improving these channels of communication, that that is not all that we want. We want the attention of Congress as well to that northern improvement which I have alluded to. We want the improvement of the St. Clair Flats, and also of that magnificent outlet, the Mississippi river. Also to commend to their special attention, the construction of a canal around the Falls of Niagara. It is for this purpose I have been induced to offer this amendment to the resolution.

The amendment to the amendment was then put and carried, and the chair announced that the question reverted to the amendment as relating to the Des-Moines, and the Fox and Wisconsin improvements.

Mr. JENNISON, of Kansas, said: I had hoped that we should assemble here as brothers, East, West, North and South. I can speak for the people of Kansas, that we join hands with you in this national enterprise, believing it a commercial and military necessity that the canal be built. The Western people have interests as well as New York, Ohio and Illinois; and while we grant there is a common interest, the people of Kansas view these matters as of national interest.

The remainder of his remarks, being foreign to the subject under discussion, are omitted.

Mr. FOOTE, of New York, moved the previous question.

Mr. KELLOGG, of Illinois, asked the mover to withdraw that motion, so that he might move to recommit the original resolution and all pending amendments to the Committee on Resolutions.

Mr. FOOTE, of New York, withdrew his motion.

Mr. KELLOGG, of Illinois, then submitted his motion, and moved the previous question.

The previous question was seconded, and, under its operation, the original resolutions and all pending amendments were recommitted to the Committee on Resolutions.

Mr. MCMANUS, of New York, submitted a resolution directing the Committee on Resolutions to embody in their report a recommendation for an appropriation of $2,000,000 for the improvement of the Hudson river between New Baltimore and the city of Troy.

The resolution was referred to the Committee on Resolutions.

Mr. ARNOLD, of Illinois, presented the following, which were referred to the Committee on Resolutions:

The representatives of the loyal States, assembled in National Convention at Chicago, desirous of cementing a closer Union, of perpetuating our Nationality forever, of providing for the common defense and promoting the general welfare of our whole country, adopt the following resolutions:

1. That we regard the enlargement of the canals between the Valley of the Mississippi and the Atlantic, as of great Military, National, and Commercial importance. We believe that such enlargement, to the capacity of passing gunboats from the Mississippi to Lake Michigan, and from the Atlantic to and from the Great Lakes, will furnish the cheapest and the most efficient means of protecting the Northern frontier, and at the same time tend greatly to promote the rapid development, and permanent Union of our whole country.

2. That the opening substantially, by the improvements proposed, of another mouth to the Mississippi on Lake Michigan, and commercially at the City of New York, so that that great national and continental highway shall discharge itself at New York via Chicago, as well as at New Orleans, is a work demanded alike by military prudence, political wisdom, and the necessities of commerce. Such a work would be not only national but continental. Every motive of sound political economy requires its early accomplishment.

3. That such a national highway should, as far as possible, be FREE, without

tolls and restrictions. We should, therefore, deprecate the placing of this grand highway in the hands of any private corporation or State. The work should be done by national credit; and, as soon as its cost is reimbursed to the national treasury, it should be made free as the lakes to the commerce of the world.

On motion, the names of Messrs. D. E. Anthony and E. G. Carr, were added to the list of delegates from the State of Kansas.

Mr. Foote, of New York, submitted a resolution that the Committee recommend to Congress to provide such ample means for canal-communications from the Mississippi to the Eastern States by lakes, and the construction or enlargement of canals, as it shall determine to be a national military necessity.

The resolution was referred to the Committee on Resolutions.

Mr. Truman, of New York, submitted a resolution declaring that a ship-canal of ten miles in length, would connect Cayuga Lake with Lake Ontario, thereby extending the line of navigation seventy-five miles to Ithaca, a point sixty miles nearer to the city of New York by railroad than is now reached by vessels on the Great Lakes, which would make a more feasible approach to the Anthracite coal region, and the iron mines of Pennsylvania, and open a capacious and safe harbor for gun-boats in the interior of the State of New York.

The resolution was referred to the Committee on Resolutions.

Mr. Walbridge, of New York. I rise to propose to read to the Committee a resolution which has already received an indorsement in the city of New York, when the people of that city met to vindicate the integrity of the constitutional government of the land; that has subsequently been indorsed by the Boards of Trade of New York city, city of Buffalo, city of Toledo, and also by the city of Chicago, when, last year, the largest mass meeting which ever was convened here, was assembled. By the construction of the seventy-three miles of canal to which this resolution refers, you would be enabled, in the vicissitudes of foreign war, to find ample and easy transit, to places ranging from 150 to any number of miles along your coast, so that in such vicissitudes we could still convey our armies and our navies and preserve our commerce.

I am convinced that, between the Mississippi and the Atlantic, there should exist a stream as broad as the Mississippi, capable of passing our ships in time of war; and it is upon that subject I have a right to speak. I believe that national exigences require that it should be performed. I submit this resolution to the Committee that they may act upon it, stating that New York desires not to discriminate between any of these great States, sympathizing as she does, with all these great movements, of whatever magnitude they may be, in this gigantic strife in which we are all engaged.

Resolved, That Congress should provide for opening the great line of interior water-communication along our Atlantic coast capable of passing our naval fleet and our commercial marine from the waters of the Roanoke, and Chesapeake Bay, to the eastern terminus of Long Island; and that the loyal States, through which the work is to be constructed, should at once open the means of internal communication, by which our gun-boats can pass from the Mississippi, by the various canals and Lakes, until they reach the Atlantic seaboard, by the most cheap and expeditious route that scientific and practical knowledge may develop.

Mr. Sabine, of Massachusetts, proposed the name of George E. Hill, of Sheffield, Mass., as a delegate.

The Convention then adjourned until three o'clock P. M.

AFTERNOON SESSION.

The President called the Convention to order at three P. M.

Mr. Foster, of Illinois, presented a letter from Mr. Benton, Auditor of the canals of the State of New York, in reference to the financial condition of those works. As a mark of respect to the high source from which it emanated, he moved, not merely that it be referred to the Committee on Business, but, also, that it be printed among the proceedings of this Convention. (Appendix, F.)

The motion was agreed to, and the communication ordered to be printed.

Mr. Ruggles, of New York, from the Committee on Resolutions, reported back the following:

The representatives of the loyal States, assembled in National Convention at Chicago, desirous of cementing a closer Union, of perpetuating our Nationality forever, of providing for the common defense, and promoting the general welfare of our whole country, adopt the following resolutions:

Resolved, That we regard the enlargement of canals between the Mississippi river and the Atlantic, with canals duly connecting the Lakes, as of great National, Military, and Commercial importance; we believe such enlargement with dimensions sufficient to pass gun-boats from the Mississippi to Lake Michigan, and from the Atlantic to and from the Great Lakes, will furnish the cheapest and most efficient means of protecting the Northern frontier, and at the same time will promote the rapid development and permanent Union of our whole country.

Resolved, That these works are demanded alike by military prudence, political wisdom, and the necessities of commerce; such works will be not only national, but continental, and their early accomplishment is required by every principle of sound political economy.

Resolved, That such national highway between the Mississippi and the Lakes, as far as practicable should be FREE, without tolls or restrictions; and we should deprecate the placing this great national thoroughfare in the hands of any private corporation, or State. The work should be accomplished by National credit, and as soon as the cost is reimbursed to the National Treasury, should be as free as the Lakes to the commerce of the world.

Mr. WASHBURNE, of Illinois, moved to amend the third resolution by adding the letter "s" to the word "highway," so as to make it read "highways."

The amendment was agreed to.

Mr. GRINNELL, of Iowa, moved to amend the first resolution by inserting before the word "enlargement" the words "construction and," so as to make it read "the construction and enlargement of canals," and to make a similar amendment in a subsequent part of the resolutions.

Mr. KELLOGG, of Illinois, suggested that those words were inserted in the draft by the Committee, and omitted simply by a mistake in copying.

The amendment was made, and the word "the" was also struck out before the word "canals."

The question was then taken on the resolutions, and they were adopted unanimously.

Mr. RUGGLES also reported the following:

Resolved, That an Executive Committee, of one from each State, be appointed by the President of this Convention, to prepare a memorial to the President, and the Congress of the United States, presenting the views of this Convention, and urging the passage of the laws necessary to carry them into full effect, with power to open such correspondence as may be expedient; and, in their discretion, to call any further conventions. Five of the members of said Committee, at any meeting duly notified by the Chairman, shall constitute a quorum.

Mr. LITTLEJOHN moved to amend, by making the resolution read "two delegates from each State." The business of that committee would be of a very important character, and there were conflicting interests in some of the States. He thought the matter would be safer in the hands of a large, than in the hands of a small, Committee.

Mr. KING differed from his colleague. He thought it safer to leave it in the hands of one gentleman from each State, who would be selected for his intelligence and knowledge of the subject.

The question was taken on the amendment, and it was disagreed to.

The question recurred on the resolution, and it was adopted.

Mr. WASHBURNE, of Illinois, moved to re-consider the vote by which the resolution was agreed to, and also moved that the motion to re-consider be laid upon the table. The latter motion was agreed to.

Gen. WALBRIDGE here took the chair as temporary presiding officer.

Mr. KELLOGG, of Illinois, from the Committee on Resolutions, reported back the following resolution:

Resolved, That the report of the survey of the Illinois and Des-Plaines rivers, executed under the authority of Messrs. Gooding and Preston, be referred to a special committee of five, to be appointed by the Chair, and to be composed of engineers eminent in their profession, who are requested to report as to the practicability of the work, and the correctness of the estimates.

A DELEGATE FROM ILLINOIS moved that it be referred to the Business Committee to be appointed by this Convention.

Mr. FOSTER, of Illinois, explained. A recent survey had been made of that portion of the Illinois improvement which starts at Joliet and ends at Chicago. A survey of the rest of the chain had heretofore been completed. The resolution only contemplated the appointment of five engineers to examine as to the correctness of those estimates.

Mr. CHAMBERLAIN, of Ohio, wished to know what objection there was to have the matter referred to the Executive Committee, and have the report made as a whole.

Mr. FOSTER explained that the official report of the Illinois-river improvement had been already submitted to Congress; but they had not an official report of the Des-Plaines portion; and he wished to have the survey verified as to its correctness.

Mr. KELLOGG, of Illinois, further explained the matter. The Committee on Naval Affairs in the House of Representatives had unjustly assailed the estimates as not being entitled to consideration. It was now proposed that there should be taken from various portions of the United States five civil engineers, who might examine these estimates and set the matter at rest.

The motion to refer to the Executive Committee was withdrawn.

Mr. RUGGLES, of New York, entered into a long explanation of the matter. As his friend (Mr. Kellogg) had just observed, the cost of this great chain of works had been outrageously exaggerated by a hostile Committee of the House of Representatives. That Committee had made a report which, he would take the liberty of asserting here, in the presence of this American multitude, was a most unworthy state-paper—a slander upon the State of Illinois, and a slander on the State of New York. They had taken measures in the State of New York to prove that report to be false and preposterous. The last Legislature had directed further surveys to show the utter and wanton extravagance of the allegations of that report, that the proposed enlargement of the locks of the Erie and Oswego canals, with the necessary alterations of their channels, would cost *twenty millions!* They had, furthermore, requested their Governor to ask the President of the United States to detail an officer of the United States to unite in the survey, so that next winter they would have the means of nailing that slander to the counter. He was, therefore, in favor of this resolution, so that Illinois might have the same advantage as New York.

A MEMBER from Wisconsin renewed the motion to refer the resolution to the Executive Committee. He should like to know why it was that that Committee could not examine this subject, and appoint, if necessary, competent surveyors to survey all those routes. Why should they attempt to commit this Convention to one route in preference to another? The Fox and Wisconsin river-improvement wanted estimates made, too. And so did others. The fair way was to refer that resolution to the Executive Committee, and let that Committee take such measures as would be just to all the routes, instead of undertaking seemingly to indorse one particular route in preference to others.

Mr. CHAMBERLAIN, of Ohio, thought it would have been better if this resolution had not have been offered at all, so that the Executive Committee might have been left perfectly untrammeled to look over the whole ground. But as one particular route had been presented, he proposed that the whole of the routes be presented in the same manner. There were half a dozen routes in the State of Ohio through which they would like to have canals constructed. Let them all have their chance. Let the Committee appoint engineers to report upon them, and they would settle down on the right channel. He did not see why gentlemen were afraid to trust the matter to the hands of the Committee. He was perfectly willing, so far as Ohio was concerned, to indorse this route to the fullest extent. At the same time, he claimed that the routes eastward from here should be properly indorsed also. And if the people of Wisconsin wanted their routes examined, let them have that privilege.

Dr. BRAINARD, of Illinois, said that so far as Illinois was concerned, they were perfectly willing that it should be made the duty of the Committee of Engineers to examine all works contemplated, and coming within the purview of the action of this Convention. But the reason why they ask this to be referred to a Committee of Engineers was, that it was conceived that engineers were the only persons capable of passing on the engineering parts of the question.

They do not propose to put the Convention to the expense of re-surveying. They had all the plans, profiles, sections and estimates; and all they asked was to have a Committee of Inquiry to examine these data. If other works were surveyed in the same way, and the surveys were brought forward, they could also be presented and passed upon by the same Committee.

A Delegate from New York thought there was no necessity for the Committee in reference to the canals of New York. An Engineer of the General Government, in conection with the State Engineer, had made a survey of the Niagara ship-canal, and it would be disrespectful to have their works revised by a Committee of Inquiry.

A Delegate from Wisconsin spoke against the appointment of the Committee of Engineers. The Fox and Wisconsin route had been surveyed, and he said they did not want the indorsement of the route, but neither did they want the implied indorsement of any other route. They expected that the Executive Committee would appoint engineers to examine all the surveys. They could show that Wisconsin could be connected with the Mississippi for $2,000,000; while it would require $20,000,000 to make the conection by the Illinois-river route.

Mr. Foster, of Illinois, hoped it would not be supposed for a moment that he had any sinister motive in introducing the resolution. He was in favor of the improvements demanded by Iowa and Wisconsin. If we could satisfy the country that the Illinois improvement would cost a fixed sum, it was information greatly to be desired, and would form the basis of future calculations. If the delegates from Wisconsin were prepared to give similar information in regard to their route, it was certainly to be desired. If he had had any idea that his resolution would have given rise to debate, he would not have introduced it, and now he was willing to withdraw it.

A Delegate from New York moved as a substitute, that all delegations be invited to procure surveys and estimates of the routes which they are in favor of, and have them submitted, with the one already provided, to the Committee of Engineers, who are to make a report on all the propositions.

The DELEGATE from Wisconsin accepted that for the motion which he had submitted.

Another DELEGATE from New York moved, as an amendment, to give to the friends of other routes the same length of time to procure their surveys and estimates, that the friends of the Illinois-river route had had.

A DELEGATE proposed to lay the whole subject on the table. The Executive Committee to be appointed would have full power over the whole subject. This Committee should not indorse any particular route.

Mr. FOSTER remarked that the question had awakened so much sectional feeling that he would now ask leave to withdraw the resolution.

The resolution was accordingly withdrawn.

A DELEGATE from New York submitted a resolution, that the friends of the various routes of communication between the Mississippi river and the Atlantic, be invited to procure surveys and estimates of the cost of the different works, and submit them to a committee of five engineers, to be appointed by the Convention.

A MEMBER. At whose expense?

The NEW YORK DELEGATE. At their own expense.

After some remarks and suggestions, the resolution was withdrawn, and thus the whole matter dropped.

Mr. S. B. RUGGLES, of New York, offered the following resolution:

Resolved, That this Convention is of opinion that the increased stimulus to be given to agriculture and commerce, by cheapening the transportation of Western products through the proposed enlarged canals, will so far increase the foreign commerce of the country, that the import duties on the return cargoes will very far exceed the interest on the cost of the proposed works, and provide a fund for its rapid reimbursement.

In asking the attention of the Convention, for a few moments, to the subject of this resolution, Mr. Ruggles expressed his fears that it might possibly interfere with the rather singular desire of a portion of the audience to listen to political harangues on the rebellion and similar topics. He would only claim for the resolution the comparative merit of being somewhat kindred to the objects for which this Canal Convention was assembled, and would seek to enforce its passage only by a brief statement of a few simple facts.

Much had been said as to the merits and necessities of various channels of national intercommunication, having for their object the military defense of the country, and also the cheapening of the transit of food to the seaboard; but, as yet, little or nothing, as to their probable cost, or the means possessed by the nation for reimbursing it.

He should not seek on this occasion to sum up the various elements of national and fiscal strength to be called into being by the construction of the proposed works—and especially and manifestly the creation and increase of large and populous, and prosperous communities, rich in every species of property and business, and able to contribute liberally in taxes and imposts to the support of the Government, but confining himself within the narrowest bounds, he should seek to specify and put his finger precisely on the very moneys with which the nation could immediately pay for the proposed works, and which would be, moreover, furnished by the works themselves.

For this purpose he begged to repeat, and he stood ready to prove the truth of the proposition contained in the resolutions now under consideration, to wit: "*that the import duties, to be paid into the national treasury upon the return cargoes, purchased by the additional amount of food to be exported, will not only exceed the interest on the cost of all the works needed to secure that increase, but will provide a fund for its rapid reimbursement.*"

I know not, said Mr. Ruggles, how to state this proposition more plainly or precisely, nor can I believe that there can be found in all this assemblage a single individual so dull or so prejudiced, as not to see and admit its truth. But if there be any such, let him now rise and show himself. Let him deny, if he can, or dare, that every thousand dollars' worth of food exported to Europe will there purchase its equivalent thousand dollars' worth of foreign products in exchange, and that this return cargo will pay into our national treasury, under the existing tariff, an average duty of at least thirty per cent., or three hundred dollars, and that, too, in gold. I pause, said Mr. Ruggles, for an answer. Does no one deny it? Can no one deny it? In discharging the official duty committed to me by my State, I have repeatedly propounded the same question, at Washington and elsewhere, to eminent legislators and statesmen, and have earnestly besought an answer, as I do now and here, from you or any of you. I never yet found the man to deny the proposition.

Some, it is true, seeking to evade its force, have skulked away, under the cry that the country was at war, and had no money but for war; that it really could not afford the means even to enrich itself—as if a period of war and lavish expenditure was not the very time and the very reason for replenishing, to the utmost, our wasting fiscal strength. As well might the farmer refuse to sow his land or seek for a crop because one of his sons had gone to the war. Political philosophers have also been found so deeply imbued with the abstractions of the Southern school, just now a little out of vogue, as to deny the constitutional authority of the Nation, thus to fill the national treasury and re-invigorate the national strength; while others again, in a spirit of mean and narrow sectionality, have thought it patriotic to excite local jealousies against the proposed highways for the national commerce, as tending to lessen the value of existing channels in other localities, and have especially asserted the impiety of attempting to compete with the Mississippi as the stream specially designed for our internal trade by God himself. But, I repeat, not one of them has yet been found hardy enough to deny the simple fact, that every thousand dollars exported in food will virtually bring back, directly or indirectly, another thousand dollars on which three hundred dollars of duty will be paid to the nation in gold.

Fellow citizens of the fertile and mighty West! Need I say, for do you not already know, that it is you, and you only, who now fiscally support the Government of this American Union? For is it not you, and only you, who now practically furnish the products for export which bring back through the returning imports that golden stream of duties by which alone the public credit is now supported, in this hour of effort and struggle? The time was when the cotton of the South contributed largely to the work of filling the treasury. Its day has passed, at least for a season. Food has taken its place. Our foreign exports now consist almost exclusively of food, or its indirect product. It is true that the tables show a moderate export of the products of the forest (but the largest portion even of that is furnished by the West,) and a considerable amount (about thirty millions) of the manufactures of the Eastern States, but the latter is only the reproduction in another form, of the food from the West, which feeds the manufacturers, and without which not a water-wheel or a spinning jenny would move. In truth, AMERICAN FOOD has now become the very substratum, the vital, vivifying principle of AMERICAN COMMERCE. The food of the vast interior, once derided as the dream of the enthusiast, no longer floats in the regions even of enlightened faith or prophecy, but now stands before the world, crystalized into a solid and unchangeable reality—an immense and unmistakable fact. It towers high, in plain sight, in the enormous masses of the cereals, and the product of animals fed by those cereals, now sent down to the ocean by the eight great Food-producing States, so richly clustered around the Lakes and the Upper Mississippi.

My friends, every subject has its one great central, dominant idea—like gravitation in the solar system. The central truth in the vast subject now before us is this, that the gigantic masses of food, which are already radiating from this unrivalled group of States, are to feed not only all the Union around it, but a large portion of the world across the Atlantic.

But let us, before proceeding to take in this whole idea, pause for a moment. Let us take in only a minor portion of the subject, and see by a few figures the influence which that minor portion alone already exerts on the fiscal condition of the nation. Last year our, and your, good city of New York, (for it is quite as much yours as ours) sent to Europe fifty millions of your cereals in grain. They purchased in return at least $50,000,000 worth of foreign merchandise, which merchandise paid in duties to our esteemed friend, Mr. Secretary CHASE, at least fifteen millions of dollars in gold. Let me ask, what would become of Mr. CHASE without you of the West, who produce this grain, and us of the East, who carry it to the seaboard and thence across the ocean? Without your broad and fertile fields, and our heavily laden canals, railways and ships, his government would not be solvent a single day. It is you, and only you, the men of the Food-producing West, that furnish "the stuff" to pay his debts. Fellow citizens, the time has come for you to take your own measure, and fully to claim your height to the last inch. But, in this vast audience, is there one to be found with head or heart large enough, or with eye clear or strong enough to discern, to survey, or to comprehend the whole marvellous future of your inevitable export of food? Who would dare attempt it? Why, last year, when you had only rudely scratched a few patches of the surface of your boundless prairies, you sent down through Lake Erie one hundred and eight millions of bushels of grain, to knock at the gates of the New York canals and nearly choke their channels; and yet it was but little more than one-fifth of the 520,000,000 bushels, the actual product, for the year, of your teeming soil.

But where shall we find an adequate measure, even for the scanty portion thus finding its way to the sea? Do you really know—are all of you fully aware—(and if you are not, you can ascertain the fact in five minutes by measuring a barrel) that these 108,000,000 of bushels, placed in a line of barrels crosswise, would span the Atlantic and reach nearly to the confines of Asia? My friends, do not become prematurely excited, or wait at least until you find that your whole yearly product of 520,000,000 of bushels thus barreled up would encircle the globe. You will thus find somewhat of an adequate measure for your crop—somewhat of a proper yard-stick with which to march into the future.

For who, in all this large assemblage, regards for a moment these 520,000,000 bushels as the full measure, or even a tithe of your product, when the whole of your 260,000,000 acres shall be brought into full and careful cultivation? True, it already exceeds the whole cereal product of the British islands, and nearly approaches that of carefully cultivated and carefully governed France; but can a man be found upon these magnificent Western waters small enough, or stupid enough to assert, that these eight great States have now reached their full maturity, have now got all their growth? What human being in his senses, not wholly idiotic, or utterly blinded by political bigotry, or lust of political power, could assert that this God-given, exuberant and all but virgin West has now reached its "culminating point"? For one, I stand awe-struck and amazed at the immeasurable prospect opening before us. I can see nothing smaller, nothing more diminutive, nothing less stupendous, than a yearly product of cereals, to be measured not, as now, by hundreds, but by thousands of millions of bushels—a result so vast, so solemn, so fraught with consequences so momentous to our nation and to the world, that I can but bow with reverential gratitude before such a wonderful manifestation of the providence of our great Creator. Never before in human history did He lay out a garden so wide-spread and fertile; never before did He provide a granary so magnificent for the use of man.

For what was ancient Sicily, the "granary of Rome," or the fertile plains of the Po, or the exuberant valley of the Nile itself, compared with this our great continental garden, pouring forth yearly volumes of food so enormous and yet so inevitably, resistlessly increasing? In view of such a power to feed our race, who will venture to depict or limit the commercial and the political destiny of this unequaled portion of the earth? Was it thus specially endowed and set aside by the Great Architect of Nations merely to feed the petty State of Illinois, great as it is, and large enough to hold a half dozen Sicilies; or the still more petty State of New York, with all its golden gates of commerce; or rocky little New England, with its thousand and one "notions" on land, and its ever "victorious industry" both on land and sea; or even the whole majestic Union of these temporarily jarring American States, soon, I trust, to be happily pacified?

No, my fellow countrymen, the manifest destiny and high office of this splendid granary, of which this Chicago of yours and of ours is the brilliant centre, stands out plain as the sun in heaven. It is unmistakably marked by the finger of God on these wide-spread lands and waters, that it is to be our special duty to feed not ourselves of this New World alone, but that venerable, moss-covered fatherland—that old father world of ours across the ocean—as the pious Grecian daughter nourished her aged sire—to carry abundant food, and with it the means of higher civilization and refinement, and that too in the truest Christian spirit, to that overcrowded but under-fed European Christendom to which we owe our common origin. Let us then come fully up to the measure of this world-wide idea. Let us, by cheapening the transit of food to our seaboard, prepare vigorously to carry out the predestined and providential arrangement of God himself to increase the happiness of man.

And now, my esteemed friends, let us make a slight descent; let us talk a little about hogs, and the glorious West as a gigantic hog-pen. I must really beg you not to laugh, for I am profoundly serious, and do earnestly assure you that the hog is a very praiseworthy, interesting, and important animal. For how, let me beg to ask, could you possibly, without his benevolent and efficient aid and co-operation, bring down the whole of these five hundred millions of bushels of grain to the sea? How could such a mountain mass of cereals, and especially of Indian corn, ever be sold or disposed of? But, thanks to the ingenuity of man and the necessity of the case, the process has been found. The crop is condensed and reduced in bulk by feeding it into an animal form more portable. The hog eats the corn, and Europe eats the hog.

4

Corn thus becomes incarnate; for what is a hog but fifteen or twenty bushels of corn on four legs?

It is among the many providential features, of which this subject is full, that a striking revolution has taken place just within the last two troubled years, in the destiny of the American hog. By a new process of curing or preparation, brought in, as I am told, from England, the animal has suddenly become extensively marketable in Europe.

Heretofore, the quadruped has passed after death into brine, obedient, perhaps, to the traditions of New England, where a pork-barrel in every family is a sacred institution. But Europe did not relish, and would not eat the hog in brine—so that a great hog-reformation is now in vigorous progress through these interior States, in packing the animal, not in brine, nor in a barrel, but in dry salt, in a light, cheap wooden box. In that shape Europe has recently consented largely to eat him. But let us ascertain precisely and statistically just how far the tickling the palate of the Old World has already advanced. In the year 1859, the exports of pork in the box (barbarously denominated "cut meats" in the official tables) were only nine millions of pounds. In round numbers they rose to twenty millions in 1860, to seventy millions in 1861, to one hundred and thirty millions in 1862, and during the present year, 1863, will probably very nearly ascend to three hundred millions of pounds. Inverting the calculation, and bringing the "cut meats" back to "hog" again, this export is equivalent to an army of one million and a half of these interesting animals, marching across the ocean. After this, will you, can you laugh at the hog?

At any rate, you will consent to be more serious when you perceive the fiscal effects of such a swinish exodus on our national treasury. These three hundred millions of pounds are worth in Europe thirty millions of dollars, sending back imports, paying in duties nine millions of dollars in gold.

Nor is this quite all. We have a little more of "the whole hog" in a fiscal point of view. The skill of our artificers in pork expresses out the very quintessence of the creature into lard, an humble element which has suddenly risen from its ancient culinary office of making cakes and greasing kitchen utensils, to the more exalted duty of illuminating houses, and oiling the millions of wheels of our locomotives, and other labor-saving machines. Not only has it literally smoothed our way to this very Convention, in this great hog-manufacturing city, but it is exerting its world-wide influence in relieving the whales within the Arctic and Antartic circles from the indefatigable pursuit of that same rock-bound, but vigorous New England.

But to descend, or rather to ascend again into figures—the foreign export of lard has so kept pace with the hog in box, that New York during the present year will send out nearly one hundred and fifty millions of pounds—worth abroad at least fifteen millions of dollars—bringing back duty-paying imports, yielding the further amount of four and a half millions of gold to the national treasury. Do you not see, my friends, how gallantly and patriotically, in Mr. CHASE's great financial struggle, the hog has come to the rescue? Respect him then, I beg, as your and our great co-operator and much esteemed associate in the great business of removing, in fact of "rooting out" every impediment to our internal commerce.

But again—this most necessary and valuable process of transmuting our vast and overwhelming supplies of Indian corn and other grains into animal forms, holds true to a large extent with the countless herds of horned cattle, so richly fed on our Western prairies, only to be hurried off to the great beef-eating cities and communities on the Atlantic. The ox, now on the dinner table in our great metropolis, was feeding but forty-eight hours before on the Mississippi; and thus equally with the hog plays his patriotic part in swelling the mighty stream of our domestic and foreign commerce.

I might proceed yet further, and but for my dread of the displeasure of our temperance friends, might trace our Indian corn onward and upward into its spiritualized condition, furnishing whisky and alcohol, by millions on millions of gallons, not only to our own thirsty countrymen, but largely to France, thence to be returned to us as genuine Cognac; another striking proof of the value of our corn, in animating and inspiriting our foreign commerce; but I will detain you no longer.

My object in these remarks has only been to vindicate the fiscal truth stated in the resolution now before you. In so doing, I have mainly sought to exhibit the imperative and solemn obligation of our National Government, promptly to exert all its power to cheapen to the utmost, the transit of this great mass of agricultural wealth to the ocean. The necessity for thus invigorating our fiscal resources is so plain and so transcendent, as to override all the prejudices and all the doubts of that pestilent school of political thinkers, who formerly exerted so pernicious an influence in our national councils. The only sensible question now is, can or cannot the nation afford thus to benefit its fiscal condition; thus to swell, and enrich its great streams of national commerce; thus to reward and encourage the industry of the American people, and especially at a moment like the present, when the hand of taxation must necessarily and permanently be laid heavily upon them? Surely a paternal government, in imposing such a burthen, would do wisely in strengthening the ability of the people to bear it.

The cost of all the works, for which the aid of the Government will probably be solicited by this Convention, will not exceed twenty-five millions, or thirty at the utmost, and even that is not asked, in money, but only in the six per cent. bonds of the Government, payable in twenty years.

The yearly interest even on thirty millions would be but $1,800,000; whereas, every $10,000,000, added to our agricultural exports, will yearly yield in return duties $3,000,000, fully meeting the interest, and amply providing a fund for rapidly reimbursing the principal. But who, with the facts now before us, and the reason that God has given him, could think for a moment of limiting that increase to ten millions? Rely upon it, that long before the twenty-year bonds shall fall due, the increase will far more probably exceed one hundred millions, if not a much larger amount. The Government bonds to be issued for the comparatively trifling sum required, will melt away like snow flakes before the rising sun, while the wonder-working channels of commerce and defense, fully paid for and exempt from debt or burthen, will remain, through the coming ages, to exert their beneficent and benignant power, advancing and securing in constantly increasing measure, the prosperity, strength, and happiness of the American people.

The resolution was adopted by the Convention by unanimous acclamation.

Mr. DRAKE, of Missouri, offered the following resolution:

Resolved, That the thanks of this Convention are due and are hereby tendered to the President, for the able, courteous, and dignified manner in which he has discharged the duties of the Chair.

The resolution was carried unanimously.

The PRESIDENT then responded as follows:

Gentlemen of the Convention: For this kind expression which you have seen fit to tender to me, I beg you will accept my undissembled and cordial thanks. In after times, whatever my lot in life, it will be to me a pleasant reminiscence that I had the honor of meeting and presiding over so large, and so intelligent a Convention, representing the substantial business and commercial interests of the country, as has here assembled. I shall carry it with me as a proud recollection in after life. I beg you to believe that, in whatever position I may be placed, however humble may be my acts, they will always be devoted to developing the resources of our mighty country, and advancing it in its material interests. You may leave New England out in the cold, if you please; but, Sir, she has a warm and generous heart, which will beat responsive to the vast and majestic West. Nay, you may sever, if you will, all her water-communications and her railroads; still, she will cling to the generous and patriotic West as she will cling to all the Union. Indeed, when I am here in your midst, I am absolutely puzzled to know how New England is to be separated from the West. The only complaint I have to make against so many of your good citizens here, is that they have deserted New England to come to you. They form a connection which nothing can sever. You are bone of our bone and flesh of our flesh. All that interests you interests us, and all your interests are ours—one interest and one destiny must alike await us.

Gentlemen, allow me to congratulate you upon the unity and harmony of your deliberations, which will exert a moral power throughout the country, and to wish you a safe and pleasant return to your homes and families.

Mr. RIDDLE. I would ask whether the Business Committee has anything before it?

Mr. WALBRIDGE. That Committee is exhausted. It has no business before it.

Mr. RIDDLE. I would suggest, then, that the Committee of one from each State is the next thing in order.

While the President was occupied in designating the Executive Committee, several distinguished delegates were called for.

Gen. JOHN COCHRANE, of New York, was introduced as one who talked very well, but who fought better.

Gen. COCHRANE said that he would rather fight than talk. The little, therefore, which he would submit to them now would be more of the military, than of the civic cast. He congratulated the Convention and the country at the auspicious conclusion of these proceedings. The Convention had avoided the rock which threatened it with danger, and had laid a solid basis for a work of military defense, without regard to commercial advantages. Indeed, there were those, himself among them, who were in great doubt how to distinguish between military defense and commercial advantages. This Convention had adopted a platform which would be copied by future political parties in this country—the great platform of progress and of commercial wealth. It was that which was to inspire them with vigor, to nourish the future gigantic growth of this great country, and to teach this whole continent that it is ours beyond the possibility of doubt, ours by the nerve and strength of our good right hand, won from the soil by the muscle of our fathers, the sturdy yeomanry of the country.

They had to teach the world this other fact, that in truth

"Westward the Course of Empire takes its way;"

That it was located on these Western plains, here in this salubrious climate, under these propitious skies, on this teeming earth. They stood here, as the great devotees and priests of that commerce, which was revolutionizing and impelling forward in its progress the destiny of this Western world. The Convention had done well. It had linked the fair East to the glowing West; and they would behold, at no distant day, that Siamese ligament—the falls of Niagara—which should connect the East with her twin sister of the West.

Gen. WALBRIDGE, of New York. I am requested by the President to submit the following resolution:

Resolved, That the liberal-minded citizens of Chicago are entitled to the best thanks of the visitors to the city, for the rich and splendid provision made for the accommodation and entertainment of this Convention.

The resolution was carried by acclamation.

The President next introduced Governor WASHBURNE, of Maine, who addressed the Convention.

The North-east State had sent her representatives here to take counsel with others in respect to the needs and rights of the West, and she rejoiced to declare her earnest and unprompted recognition of them all. Whatever else might be said of her, she was not mean nor narrow; but in her principles, feelings and instincts, was, as she had shown on many occasions, by her works rather than her words—for she was not cunning in phrases of self-commendation—broad and catholic. Upon questions of river and harbor-improvement, and on other measures proposed in Congress for the advantage of the West, her votes had generally, if not always, been such as the West had given herself. To you, he said, she has given her hopes and good wishes, and thousands of her men and women, whose industry, enterprise, and intelligence, have neither been unnoticed nor unfelt in the marvelous conquests which you have made; she has given you patriotic men, like those you have heard from in this war, whether led by her own Berrys, Howards and Jamesons in the East, or by the Grants, Rosecranses, and others of the West, whose deeds belong to history, and whose names are written on the red-leaved tablets of the heart, ineffaceable forever.

Look, Mr. President, at the map of the United States, and you will observe that Maine has two hundred and fifty miles of direct sea-frontage; and, following the line of bays, harbors, creeks and inlets, that her shores are washed by tide waters to the extent of more than three thousand miles. She has more deep, safe, capacious and accessible harbors than are to be found from the Delaware to the Del Norte. More than half her population live within an hour's ride of navigable waters; and so you will not be surprised that she builds ships and sails them—that her men, as sailors, masters, merchants, go everywhere. There is not an island in the farthest sea but has been visited by the sons of Maine. Could such a people be narrow or illiberal? Could they, with their experience, help seeing that a broad and generous policy was the wisest? Could the "Sun-rise State," with its facilities for commerce, its healthy climate, and manufacturing capabilities, doubt that whatever benefited the West, would help her? Nothing could be done legitimately to promote the welfare of the West, that would not be of ultimate advantage to New England and the entire East.

He said there were certain things to which the West was pre-eminently adapted, and to which it would be most profitable to confine herself, not altogether, but in the main. She could not afford to be a common carrier or a manufacturer; she could do better as an agriculturist. To her was granted the peculiar favor of filling the most honorable occupation among men,—not that hers would be an exclusively agricultural people, for they would not; they would be more or less engaged in manufactures and commerce, also; but their leading, characteristic employment would be agriculture. The East, with its less fertile soil, could not compete with the West in the products of the field, and while to some extent her people would be farmers, that would not be their chief business. They could not compete with the possessors of the fat acres of the West, and so would, of necessity, engage in other employments. Their soil, climate and place on the continent, had appointed them the carriers and manufacturers of the country, not exclusively indeed, but more than any other section. And was it not for the advantage of the East that she should be the carrier and manufacturer for fifty millions of people in the West, instead of ten? That her trade would be increased five-fold, and with this, her population and wealth? Nothing, he would repeat, could be done to help the West, that was not, also, for the advantage of the East. Enlarge the canal-communications from the Mississippi to the Atlantic, as you propose, and the expense of transportation of a barrel of flour, a bushel of corn, or box of pork, would be reduced to one-half the present cost. The Western producer would undoubtedly be benefited more than any other class, but the Eastern consumer would also be benefited—he would not purchase a barrel of flour without realizing a direct advantage,—but the increased trade and expanded markets, which these facilities postulate, make him even more interested in their completion, than does the simple fact of the reduced cost of the Western staples of which he is so large a purchaser. The nearer production and consumption—agriculture and commerce—are brought to each other, the better for both.

Man's business is not merely to appropriate what nature has provided, but to work with nature, and, by such co-operation, to develop the resources and increase the material wealth of a country. The outlets of the West are South and East; and nature has not more distinctly marked the channel of the Mississippi for one of them than she has indicated another, or others, towards the East. The relative position of the continents, the oceans, winds, currents, climate, population, have an important influence in channeling the courses of trade; and they have determined that the great volume of Western commerce shall flow Eastward. Nature is the true economist. She wastes no power; and so when she opened the highway of the Mississippi from the far North to the Gulf of Mexico, she declared that it was a natural and convenient channel of trade and intercourse; but not absolutely indispensable, not *so* necessary, that the enterprise and capital of men would be certain to supplement what she had left unfinished: and, when the way by the lakes and rivers to the Atlantic was but half opened, she knew that its completion was so important, so imperiously demanded by the necessities of both East and West, that the work might be safely left to be disposed of by human skill and power.

It has been suggested that the present is an inauspicious time for this enterprise. I cannot think so: on the other hand, I regard it as the appointed and accepted time for this grand national work. In a higher and deeper sense than men have generally seen, this civil war is a war for the sake of the Union. The rebellion for the destruction of the Union was needed, perhaps, to make it eternal—the union of East and West, as well as of North and South. It has secured already a Pacific railroad, which, without it, might have been delayed until a commercial, if not a political, disunion should have been accomplished. It has summoned this noble Convention, whose united voice to-day, representing every loyal State from Kentucky to Maine, has settled the question of a ship-canal from the Mississippi to the Atlantic; indeed, has worked a power that directed in one effort, shall cleave the solid earth from the Great River to the Lakes, and from the Lakes to the Ocean! Be sure, the contract, implied if not expressed, when the rebellion was commenced, will not be fully executed until all obstacles in the way of a 'more perfect union' shall have been overcome, until all sections shall be united in indissoluble bonds, social, political and commercial.

Tell me not that the country is engaged in a gigantic civil war, and has neither time nor means to do more than save the Union of North and South. We have the means and power for what is demanded; more power for the accomplishment of all things required for the security of the national life than for a part of them. In a war for the Union, let us provide against the contingency of future wars for the Union, and for adequate defenses against foreign powers. Now, if ever, we should prepare to defend the Lakes of the West, and the country wherever exposed; and he would say that some of the defenses of the West and of the country, most practicable and indispensable, will be required upon that wedge separating the upper and lower British Provinces, known as the State of Maine. The West will not wait to be defended only at the gates of Detroit and Chicago. No, the means and the opportunity for these great works of commercial, political, and military necessity, are not wanting. So far, we have carried on the war almost without an effort, with one hand only; nay, with not so much as that; we have not employed as yet even the little finger of our power. See how the country has been going on all this time—planting, harvesting, ship-building, manufacturing, trading, building cities like this modern wonder of yours, the Queen of the Prairie and the Lakes, as in times of profound peace and high material prosperity. A gentleman told us this morning that the country was "staggering" under the weight of the rebellion. Do all these badges of prosperity, these signs of unexhausted and almost untouched power, indicate anything of the kind? *Staggering!* The granite peak of Katahdin shall sooner move from its everlasting foundation, than this nation, this mighty, free people, *stagger* beneath the blows of a slaveholders' rebellion! Not only does the nation not stagger in the war, but its vast means, and the intelligent patriotism and unconquerable devotion of the people, should assure us that it will be ended with no material diminution of its strength and resources. It is weak and impious to suggest doubts of the result. Shall it be acknowledged that the great Republic is a failure, and now to pass away? Shall we flout God's providence, which has written on so many pages its necessity, and grandeur, and beneficence? Shall we not believe that it is to be saved, when we feel that it ought to be? When we see that civilization needs it, that both hemispheres need it, that the humbling of the world's despotisms and the consecrated revenges of Freedom need it? Shall we forget the circumstances of its history—the discovery of the continent, the settlement of North America, the marvellous growth of the United States? Shall we believe that the War of Independence was in fact a step backward, and its final cause the disappointment of the friends of liberty?—that God's gift of Washington was in vain?—that that miracle of human wisdom, as it has appeared to us, the Constitution of the United States, was but a wretched mockery?—that this uprising of slavery (the chief if not the only enemy of the Government) against itself, is of no significance?—that all the signs of promise, and all the hopes for man in the New World, are barren and fruitless? Oh, no! rather let us accept as prophetic the words of the good Bishop Berkeley, just quoted by my friend from New York (Gen. Cochrane):

> "Westward the Course of Empire takes its way—
> The first four acts already past;
> A fifth shall close the drama with the day:
> *Time's noblest offspring is the last.*"

Mr. Walbridge introduced a series of patriotic resolutions, which, not coming within the purview of the resolutions adopted by the Convention relating to business, are omitted.

The President announced the following as members of the Executive Committee:

Illinois,	Mr. I. N. Arnold.
Indiana,	" Geo. W. Julian.
Kentucky,	" Samuel L. Casey.
Massachusetts,	" Henry L. Dawes.
Maine,	" T. C. Hersey.
New Hampshire,	" Thos. M. Edwards.
Vermont,	" Justin L. Morrill.

Michigan,	Mr. D. Stewart.
Rhode Island,	" R. J. Arnold.
Connecticut,	" C. Day.
Missouri,	" T. J. Homer.
Ohio,	" P. Chamberlain.
Iowa,	" R. P. Hill.
New Jersey,	" Ezra Nye.
New York,	" A. A. Low.
Minnesota,	" Robert Blakeley.
Kansas,	" D. R. Anthony.
Wisconsin,	" James T. Lewis.
California,	" J. A. McDougall.

ADJOURNMENT.

Mr. KING, of New York. I move that the Convention now adjourn *sine die.*

The motion was carried, and the PRESIDENT declared the National Convention adjourned *sine die.*

PROCEEDINGS OF THE EXECUTIVE COMMITTEE.

Pursuant to previous notice, the Committee appointed to memorialize the President and Congress, met at the St. Nicholas, New York, July 2, at 12 o'clock M., when Mr. ARNOLD, as temporary Chairman, called the Committee to order.

Present, Mr. Arnold, of Illinois.
" Edwards, of New Hampshire.
" Hersey, of Maine.
" Low, of New York.
" Nye, of New Jersey.
" Hill, of Iowa.
" Chamberlain, of Ohio.

On motion of Mr. CHAMBERLAIN, it was voted, that a President and Vice President be appointed,—and Mr. ARNOLD and Mr. LOW were respectively selected.

On motion of Mr. LOW, of New York, Mr. FOSTER, of Illinois, was designated as Secretary.

Mr. HILL, of Iowa, moved that a sub-committee of three be appointed to draft a memorial, to which Mr. CHAMBERLAIN offered an amendment, which was accepted by the mover, that the President and Vice President be added to the Committee. The motion was carried, with the additional amendment that the Committee be appointed by the Chair.

Mr. EDWARDS submitted the following resolution:

Resolved, That an enlargement of the channels of water-communication between the West and the East, necessary to the protection of the commerce of the Lakes, and the cities and villages on their borders, is no less important to the business interests of the East than to those of the West; and that the East, by the almost undivided votes of its delegations in Congress, has pledged itself to co-operate in securing the aid of the General Government to any proper measures for the accomplishment of this great national object.

Which, after some discussion, was unanimously adopted.

Mr. HILL moved that two additional names be added to the Committee on the Memorial, which was carried.

The Chair then announced the following as the Committee:

Mr. I. N. Arnold, *ex officio*,	Chicago, Illinois.
" A. A. Low,	"	New York city.
" R. P. Hill,		Davenport, Iowa.
" H. L. Dawes,		North Adams, Massachusetts.
" T. J. Homer,		St. Louis, Missouri.
" P. Chamberlain,		Cleveland, Ohio.
" Thos. M. Edwards,		Keene, New Hampshire.

On motion of Mr. Low, it was voted, that when this Committee adjourn, they adjourn to meet at the call of the Sub-Committee.

Mr. Low submitted the following resolution, which was adopted:

Resolved, That said Sub-Committee be authorized to invite from Boards of Trade, Chambers of Commerce, and other public bodies interested in the enlargement of the existing lines of communication between the East and the West, and in any addition thereto, such information as may bear upon the contemplated work of intercommunication, whether of a military or commercial character.

The Committee then adjourned, subject to the call of the President.

APPENDIX.

[A.]

NECESSITY OF A SHIP-CANAL BETWEEN THE EAST AND WEST.

"The Congress shall have power to levy taxes, duties, imposts and excises—to provide for the COMMON DEFENSE, and promote the GENERAL WELFARE of the United States."—CONSTITUTION.

The Committee, appointed to collect statistics as to the importance of uniting the waters of the Mississippi with those of the Atlantic by a Ship-Canal, have discharged the duties imposed upon them, and submit the following

REPORT.

Two schemes for the accomplishment of this object have been brought prominently before the country, and failed, only by a few votes, to receive the sanction of the Thirty-Seventh Congress.

1. To make a slack-water navigation of the Illinois and Des Plaines rivers, and to enlarge the present Illinois and Michigan canal to such dimensions as shall admit of the passage of gun-boats, and of the largest class of Mississippi steamers, to the Lakes.

2. To enlarge the locks of the Erie and Oswego canals of New York, to such dimensions as shall pass an iron-clad gun-boat 25 feet wide and 200 feet long, and drawing not less than 6 feet and 6 inches water.

The cost of construction of the first will be about $13,500,000, and that of the second, $3,500,000;—detailed estimates of which will be presented to the Convention.

In devising an extensive system of internal communication, it is of the highest importance to inquire into the resources of the region which it shall traverse; its topography, soil and climate; its population, products of industry and internal commerce; and its past and prospective growth;—all are elements to be taken into consideration to enable us to form an intelligible opinion as to the necessity of executing such works, and the scale of magnitude on which they should be projected.

PHYSICAL CHARACTER OF THE MISSISSIPPI BASIN.

The Valley of the Mississippi, bounded on the one hand by the Rocky mountains, and on the other by the Alleghanies, embraces a drainage area of 1,244,000 square miles, which is more than one-half of the entire area of the United States. The Upper Mississippi Valley is composed of three subordinate basins, whose respective dimensions are as follows:

The Ohio basin	214,000	square miles.
The Upper Mississippi	169,000	" "
The Missouri	518,000	" "
Making a total of	901,000	" "

Its navigable rivers are as follows:

Missouri, to near the Great Falls	3,150	miles.
Missouri, above Great Falls to Three Forks	150	"
Osage, to Osceola	200	"
Kansas	100	"
Big Sioux	75	"
Yellow-stone	800	"
Upper Mississippi, to St. Paul	658	"
St. Anthony, to Sauk Rapids	80	"
Above Little Falls, to Pokegima Falls	250	"
Minnesota, to Patterson's Rapids	295	"
St. Croix, to St. Croix Falls	60	"
Illinois, to La Salle	220	"
Ohio, to Pittsburgh	975	"
Monongahela, to Geneva (slack-water, 4 locks,)	91	"
Muskingum, to Dresden " 8 "	100	"
Green River, to Bowling Green " 5 "	186	"
Kentucky, to Brooklyn " 5 "	117	"
Kanawha, to Gauley Bridge	100	"
Wabash, to Lafayette	335	"
Salt, to Shepherdsville	30	"
Sandy, to Louisa	25	"
Tennessee, to Muscle Shoals	600	"
Cumberland, to Burkesville	370	"
Total navigation	8,967	"

NOTE.—Steamboats have ascended the Des-Moines to Des-Moines City, Iowa river to Iowa City, Cedar river to Cedar Rapids, and the Maquoketa to Maquoketa City, but only during temporary floods.

It would thus appear that the internal navigation of the Upper Mississippi Valley is about 9.000 miles in extent; but, during the summer months, even through the main channels, it becomes precarious, and at times is practically suspended.

The Mississippi Valley, viewed as a whole, may be regarded as one great plain between two diverging coast ranges, elevated from 400 to 800 feet above the sea. St. Paul, the head of the navigation of the Mississippi, is 800 feet above the ocean; Pittsburgh, at the junction of the Monongahela and Alleghany, forming the Ohio, 699 feet; Lake Superior on the north, 600 feet; but the water-shed on the west, at South Pass, rises to nearly 7,500 feet.

It is traversed by no mountain ranges, but the surface swells into hills and ridges, and is diversified by forest and prairie. Leaving out the sterile portions west of the Missouri, the soil is almost uniformly fertile, easily cultivated, and yields an abundant return. The climate is healthy and invigorating, and altogether the region is the most attractive for immigration of any portion of the earth.

PHYSICAL CHARACTER OF THE ST. LAWRENCE.

The sources of the Mississippi on the east interlock with those of the St. Lawrence, which, with its associated lakes and rivers, presents a system of water-communication of nearly equal extent and grandeur.

TABLE SHOWING THE DIMENSIONS OF THE FIVE GREAT AMERICAN LAKES.

LAKES.	Greatest length.	Greatest breadth.	Height above sea.	Area in square miles.
	MILES.	MILES.	FEET.	
Superior	355	160	600	32,000
Michigan	265	80	576	22,000
Huron	210	160	574	20,400
Erie	210	45	560	9,600
Ontario	160	45	235	6,300
TOTAL	....	....	....	90,300

The entire area drained by these lakes is estimated at 335,515 square miles, and their shore lines are nearly 5,000 miles in extent, while those of the Atlantic are but 3,500.

These rivers are as diverse in character as in direction. The

Mississippi is the longer, but the St. Lawrence discharges the greater volume of water; the one abounds in difficult rapids, the other in stupendous cataracts; the one is subject to great fluctuations, the other preserves an almost unvarying level; the waters of the one are turbid, those of the other possess an almost crystal purity; the one affords few lake-like expansions, the other swells into vast inland seas. Both have become the great highways of commerce, enriching the regions through which they flow, and supplying the inhabitants with the varied products of distant climes. (*Foster and Whitney's Report on Lake Superior.*)

The commerce of these Lakes, whose annual value reaches $450,000,000—more than twice the external commerce of the whole country—is carried on by a fleet of 1,643 vessels, of the following classes:—

	No.	Tonnage.	Value.
Steamers	143	53,522	$2,190,300
Propellers	254	70,253	3,573,300
Barks	74	33,203	982,900
Brigs	85	24,831	526,200
Schooners	1,068	227,831	5,955,550
Sloops	16	667	12,770
Barges	3	3,719	17,000
TOTALS	1,643	413,026	$13,257,020

The following are the distances of some of the commercial routes, taking Chicago as the initial point:

Chicago to Fond du Lac Superior	900	miles.
" " Georgian Bay	650	"
" " Buffalo	950	"
" " Gulf of St. Lawrence	1,950	"

PROGRESS OF DEVELOPMENT.

The first colony of English extraction, planted in the territory of the Upper Mississippi, was in 1788—just seventy-five years ago—at Marietta, within the present limits of Ohio. This was the origin of that spirit of colonization, which, within the lifetime of many living men, has peopled this region with nine millions of human beings; has subdued and brought under cultivation, an area greater than that of all the cultivated lands of the British Empire; has connected the principal commercial points with a net-work of railways more than eleven thousand miles in extent; and has built up a domestic industry, the value of whose annual product is in excess of three hundred and fifty millions of dollars. Out of this territory have been carved not less than nine States, which are indissolubly linked together by a similarity of conditions in soil and climate, and by the geographical features of the country. They have already received the appellation of the "FOOD-PRODUCING" States;—an appellation which they are destined to retain for all time.

The rivers and the lakes, which water this region, offer the most magnificent system of internal communication to be found on the surface of the earth. No mountain barriers interpose to divide the people into hostile clans, or divert the great currents of trade in their flow to the markets of the world. With a soil sufficiently rich in organic matter for fifty successive crops; with almost boundless fields of coal, stored away for future use; with vast deposits of the useful ores, and the precious metals, on the rim of the great basin; and with a climate most favorable to the development of human energy; it is impossible for the mind, even in its most daring speculations, to assign limits to the growth of the North-West. When all of these elements of wealth, now in a crude state, shall have been fully developed, there will be an exhibition of human power and greatness such as no other people ever attained.

The subjoined table (A.), compiled from the Census returns of the United States, exhibits the progress of population, as well as of cultivation in these States, from 1800 to 1860; and it will be perceived that, during this period, in both these respects, the increase has been each decade about two-fold.

The appended table (B.), also compiled from the Census returns, shows that the increase in agricultural products and in domestic animals has been in about the same proportions. Comparing the whole superficial contents of these States with the portions cultivated, it will be seen that only about 15½ per cent. of the surface has been subdued; and, if population and cultivation increase in the same ratio in the future, as they have in the past, before the lapse of another decade there will be collected annually, on the borders of the Great Lakes, more than 200,000,000 bushels of cereals for exportation, giving employment to a fleet of more than 3,000 vessels, and requiring avenues of more than twice the capacity of existing ones.

[A.]

TABLE, SHOWING THE INCREASE OF POPULATION, AND OF THE NUMBER OF ACRES OF IMPROVED LAND IN THE STATES NORTH-WEST OF THE OHIO RIVER, AND THE UPPER MISSISSIPPI BASIN, FROM 1800 TO 1860.

STATES.	Area of Square Miles.	1800.		1810.		1820.		1830.		1840.		1850.		1860.	
		Population.	Improved Land.	Population.	Improved Land.	Population.	Improved Land.	Population.	Improved Land.	Population.	Improved Land.	Population.	Improved Land.	Population.	Improved Land.
Ohio........	39,964	45,365	225,675	230,760	225,675	581,295	2,892,456	937,903	4,665,000	1,519,467	7,558,750	1,980,329	9,851,493	2,339,511	12,665,587
Michigan.	56,243	......	....	4,762	....	8,765	....	31,639	....	212,267	....	397,654	1,929,110	749,113	3,419,861
Indiana	33,809	4,875	24,890	24,520	125,530	147,178	751,445	343,031	1,751,409	685,866	3,485,729	988,416	5,046,543	1,350,428	8,161,717
Illinois	55,405	......	....	12,282	72,692	55,162	325,272	157,445	931,860	476,183	2,818,373	851,470	5,039,545	1,711,951	13,251,473
Missouri...	67,380	......	....	20,845	89,805	66,557	286,870	140,455	605,117	383,702	1,653,001	682,044	2,938,425	1,182,012	6,246,871
Iowa	55,045	......	....	....	....	....	....	....	....	43,112	184,969	192,214	824,682	674,918	3,780,253
Wisconsin ..	53,924	......	....	....	....	....	....	....	....	30,945	105,930	305,391	1,045,499	775,881	3,746,036
Minnesota..	83,531	......	....	....	....	....	....	....	....	....	....	6,077	5,035	172,123	554,397
Kansas	80,000	......	....	....	....	....	....	....	....	....	....	....	....	107,206	372,855
TOTALS...	525,301	50,240	250,565	293,169	513,702	858,957	4,256,043	1,610,473	7,953,386	3,350,542	15,806,752	5,403,595	26,680,332	9,063,143	52,199,050

[B.]

STATEMENT, SHOWING THE INCREASE IN SOME OF THE PRODUCTS OF AGRICULTURE IN THE EIGHT GRAIN-GROWING STATES, FOR TEN YEARS, ENDING IN 1860.

STATES.	WHEAT, bushels.		CORN, bushels.		OATS, bushels.		RYE, bushels.		BARLEY, bushels.		SWINE, head.		CATTLE, head.	
	1850	1860	1850	1860	1850	1860	1850	1860	1850	1860	1850	1860	1850	1860
Ohio	14,487,351	14,582,570	59,078,695	70,637,140	13,472,742	15,479,133	425,918	656,146	354,358	1,601,082	1,964,770	2,175,623	1,358,947	1,657,850
Indiana ...	6,214,458	15,219,120	52,964,363	69,641,591	5,655,014	5,028,755	78,792	400,226	45,483	296,374	2,263,776	2,498,528	714,666	1,170,005
Illinois	9,414,575	24,159,500	57,646,984	115,296,779	10,087,241	15,336,072	83,364	961,322	110,795	1,175,651	1,915,907	2,279,722	912,036	1,505,581
Michigan ..	4,925,889	8,313,185	5,641,420	12,152,110	2,866,056	4,073,098	105,871	497,197	75,249	305,914	205,847	374,664	274,497	534,267
Wisconsin .	4,286,131	15,812,625	1,988,979	7,565,290	3,414,672	11,059,270	81,253	888,534	209,692	678,992	159,276	333,957	183,433	512,866
Minnesota.	1,401	2,195,812	16,725	2,987,570	30,582	2,202,050	125	124,259	1,216	125,130	734	101,252	2,002	119,003
Iowa	1,530,581	8,433,205	8,656,799	41,116,994	1,524,345	5,879,653	19,916	176,055	25,093	454,116	323,247	921,161	186,621	536,254
Missouri...	2,981,652	4,227,586	36,214,537	72,892,157	5,278,079	3,680,870	44,268	293,262	9,631	228,502	1,702,625	2,354,425	791,510	1,168,984
TOTALS..	43,842,038	89,293,603	222,208,502	392,289,631	42,328,731	62,738,901	739,507	3,997,001	831,517	4,865,761	8,536,182	11,039,332	4,373,712	7,204,810

Here is a gross sum of more than 550,000,000 bushels of cereals, the product of the eight Food-producing States for the year 1859, based on a crop which was nearly one-third deficient, as contrasted with those of 1860 and 1861.

To convey an adequate idea of the motive power required to distribute this prodigious mass, in its crude state, it may be stated that it would employ more than 64,400 locomotives, each hauling 8,500 bushels; and, if required to deposit their freight at a given depot, a train must arrive oftener than once in seven minutes, by day and by night, throughout every working day of the year.

After feeding the existing population of those States, there remains a surplus of more than 500,000,000 of bushels, to be used as seed for future crops, as food for the domestic animals, and for exportation, either in a crude state, or in a concentrated form, as beef, pork, lard, oil, whisky, etc., etc.

As an evidence of the increase of agricultural products since 1859, consequent on improved crops, and an enlarged area of cultivation, your Committee would direct attention to the provision-trade of Chicago for the last four years.

TABLE, SHOWING THE RECEIPTS AT CHICAGO OF THE ARTICLES NAMED FOR THE YEARS 1859–62.

ARTICLES.	1859.	1860.	1861.	1862.
Flour, barrels.........	726,321	713,348	1,479,284	1,666,391
Wheat, bushels.......	8,060,766	14,427,083	17,385,002	13,978,116
Corn, "	5,401,870	15,262,394	26,369,989	29,574,328
Oats, "	1,757,696	2,198,889	2,067,018	4,688,722
Rye, "	231,514	318,976	490,989	1,038,825
Barley, "	652,696	617,619	457,589	872,053
Hogs................	271,204	392,864	675,902	1,348,890
Cattle...............	111,694	177,101	204,579	209,655

Thus, the increase in cereals has been 196 per cent.; in hogs, 400 per cent.; and in cattle, 87 per cent.

Results equally marked are shown by the returns of the other lake-ports.

BLOCKADE OF THE MISSISSIPPI.

It may be said that this is the result of the blockade of the Mississippi, and that, so soon as that blockade is raised, a considerable portion of these products will seek an outlet through that channel. This is a mistaken idea, which a brief reference to the statistics of trade will entirely dispel.

The Committee of the Chicago Board of Trade, in a recent report, say:

"In the early settlement of the West, the Mississippi was the only outlet for the products of the country; but the opening of the New York and Canadian canals, and of not less than five trunk railways between the East and West, has rendered the free navigation of the Mississippi a matter of secondary importance.

"The heated waters of a tropical sea, destructive to most of our articles of export; a malarious climate, shunned by every Northerner for at least one-half of the year; and a detour in the voyage of over 3,000 miles in a direct line to the markets of the world;—these considerations have been sufficiently powerful to divert the great flow of animal and vegetable food from the South to the East. Up to 1860, the West found a local market for an inconsiderable portion of her bread-stuffs and provisions in the South; but, after supplying this local demand, the amount which was exported from New Orleans was insignificant, hardly exceeding two millions of dollars per annum."

The annual report of the Secretary of the Treasury, for the year ending August 31, 1860, shows the amount of bread-stuffs and provisions exported to foreign countries from New Orleans and New York respectively, as follows:

	From New Orleans.	From New York.
Wheat, bushels..........................	2,189	1,880,908
Wheat Flour, barrels....................	80,541	1,187,200
Indian Corn, bushels....................	224,382	1,580,014
Indian Meal, barrels....................	158	86,073
Pork, barrels...........................	4,250	109,379
Hams and Bacon, pounds..................	890,230	16,161,749

The total receipts of grain of all kinds, at that port, in no single year exceeded 14,500,000 bushels, either for exportation or consumption in the interior, which are about the receipts at Milwaukee, or Toledo. In 1859–60, the receipts were as follows:

FLOUR. bbls.	WHEAT. sacks and bbls.	CORN. sacks and bbls.	OATS. sacks and bbls.
965,860	339,348	1,722,637	659,550

These facts show conclusively that, with the navigation of the Mississippi unobstructed, the great mass of Western exports would flow through other channels.

PRODUCT OF BREAD-STUFFS FOR EXPORTATION.

The amount of cereals, which, in 1862, flowed out of the Upper Mississippi Valley and the region of the Lakes, *en route* for the sea-board, was, according to the Buffalo Trade Report, 136,329,542 bushels, which were respectively forwarded from the following points:

STATEMENT SHOWING THE SHIPMENT OF CEREALS FOR 1862.

PLACES.	FLOUR. BBLS.	WHEAT. BUSH.	CORN. BUSH.	OTHER GRAIN. BUSH.
W. Terminus B. & O. R. R.*...	690,000	....	....	550,000
" Pennsylvania Central	890,696	....	....	1,622,893
Dunkirk	1,095,365	112,061	149,654	10,173
Suspension Bridge*...........	875,000	....	....	2,750,000
Buffalo......	2,846,022	30,435,831	24,288,627	3,849,620
Oswego	235,382	10,982,132	4,528,962	1,467,823
Cape Vincent	48,576	316,403	249,369	49,047
Ogdensburg	576,394	689,930	1,120,176	18,865
Montreal....................	1,101,475	8,012,773	2,649,136	519,896
Rochester*	1,000	150,000	....	6,622
* Estimated.				
TOTALS	8,359,910	50,699,130	32,985,923	10,844,939
GRAND TOTAL, (Flour reduced to bushels)..................				136,329,542

SHIPMENTS OF CEREALS FROM FOUR LAKE-PORTS, IN 1862.

PLACES.	FLOUR. BBLS.	WHEAT. BUSH.	CORN. BUSH.	OTHER GRAIN. BUSH.
Chicago....................	1,739,849	18,808,898	29,452,610	4,516,357
Milwaukee	711,405	14,915,680	9,489	250,292
Toledo*	1,261,291	9,314,491	3,781,634	....
Detroit†	998,535	3,278,033	310,618	122,109
TOTALS....................	4,711,080	41,317,102	33,554,351	4,888,758
GRAND TOTAL, (Flour reduced to bushels)..................				103,315,611

* Amount received from Chicago deducted.
† Amount received from Chicago and Milwaukee deducted.

The mining population of Lake Superior absorb not less than 150,000 bushels of cereals, which do not appear in the above tables, and which will account for the discrepancies between the amounts shipped from the initial points, and the amounts forwarded from the secondary points. These tables are illustrative, as showing that, in this great grain-movement, the four lake-ports furnish more than fifty per cent. of all the flour, more than eighty per cent. of all the wheat, and more than seventy-five per cent. of the cereals of all kinds; while Chicago and Toledo together furnish more corn than finds its way eastward through all these avenues, and Chicago alone contributes more than forty per cent. of the whole gross product.

These statistics show to what gigantic proportions the grain-trade of the North-West—the growth of less than a quarter of a century—has attained. The first shipment of grain from Chicago was made by one of this Committee in 1838; but the earliest bill of lading preserved bears date Oct. 8, 1839, and calls for 1,678 bushels of wheat, to be delivered at Black Rock.

PROVISION-TRADE.

The provision-trade has assumed dimensions equally important. The following returns of the pork-packing in the North-West are taken from the Cincinnati Price Current, showing the number of hogs slaughtered, as well as forwarded:

	1861–2.	1862–3.
Ohio	791,099	981,683
Indiana	495,298	587,528
Illinois	835,881	1,484,834
Iowa	205,188	403,899
Missouri	138,766	284,011
Wisconsin	100,556	196,745
TOTALS	2,566,788	3,938,700
Excess over preceding year		371,912
Aggregate weight in lbs	606,788,684	854,697,900

The number of hogs forwarded by the

	1861–2.	1862–3.
New York and Erie Railroad	124,792	136,007
Pennsylvania Railroad	205,103	171,490
	329,875	307,503

No returns of the number sent through Canada, or delivered at Buffalo, have as yet been received.

The Committee have not complete returns of the extent of beef-packing in the North-West. The shipments of cattle through one avenue alone—Detroit—amounted last year to 75,964.

CORN-CROP.

But the great crop of the North-West is that of maize, or Indian corn, the yearly product of which is now not less than 500,000,000 bushels. It is easily cultivated, and yields an almost unfailing return. It is the cheapest food for domestic animals, and in a concentrated form, like beef, pork, lard, alcohol, and whisky, will bear transportation to every quarter of the world. In a crude state, it is a commodity so bulky and perishable that, loaded with the existing rates of transportation, the prairie farmer often finds it more profitable to consume it for fuel than to ship it to the sea-board. That which is retailed to the New England operative at 60 cents per bushel, nets to him less than 9 cents—the difference being used up in freights and commissions. The consequence is, that only about five per cent. of this cereal, in its crude state, reaches the sea-board.

Estimating the future by the past, it is impossible to assign limits to the productive power of the North-West. That power will keep pace with the world's demand for cheap bread—a demand always craving but never satisfied. Hostile legislation may undertake to confine its passage to particular channels, and interested parties to levy extortionate charges on its transit; but the reciprocal interest of producer and consumer will be sufficiently powerful to sweep away all such obstacles. The universal sentiment of mankind, as well as the dictates of a sound political economy, demands that products of such vital necessity to the race shall be incumbered with the least possible restraints.

MINERAL RESOURCES — LAKE SUPERIOR MINING REGION.

Prior to 1845, Lake Superior was regarded almost as a *mare clausum;*—one or two vessels in the employment of the British and American Fur Companies being the only ones whose canvas whitened those magnificent waters. The trade of that region, now estimated at $22,000,000 per annum, requires about 200 vessels for its transaction.

The copper-mining of this region has become one of the great industrial interests of the country, giving employment to probably 10,000 miners, and yielding an annual product which goes far to supply the wants for home consumption. The native metal—for under this form it is almost exclusively found—yields a copper-sheathing, which, for purity and tenacity, is far superior to any foreign product.

The cupriferous belt extends, on the southern shore of that lake, from the head of Keweenaw Point to beyond the Ontonagon—the productive portion being about 100 miles in length, and from 2 to 10 miles in width.

The following statement shows the annual yield, in tons, of the mines, from the commencement of mining operations up to the present year:

AGGREGATE SHIPMENTS OF COPPER FROM LAKE SUPERIOR, FROM 1845 TO 1862.

	Tons. lbs.	Value.
Shipments in 1845............	.1300	$290
" 1846............	29.	2,619
" 1847............	239.	107,550
" 1848............	516.	206,400
" 1849............	750.	301,200
" 1850............	640.	266,000
" 1851............	872.	348,800
" 1852............	887.	300,450
" 1853............	1,452.	508,200
" 1854............	2,300.	805,000
" 1855............	3,196.	1,437,000
" 1856............	5,726.	2,400,100
" 1857............	5,759.	2,015,650
" 1858............	5,896.	1,610,000
" 1859............	6,041.	1,932,000
" 1860............	8,614.	2,520,000
" 1861............	10,337.	3,180,000
" 1862............	10,000.	4,000,000

IRON ORES.

The ores occur in mountain masses, sufficient to furnish an unlimited quantity of the purest iron for all time. They occupy a belt from six to twenty-five miles wide, and extend from about the parallel of Chocolate river 150 miles west, into Wisconsin. The nearest point at which these ores approach Lake Superior is south of Marquette, distant twelve miles. A railroad has been constructed sixteen miles in length, so as to intersect three of these great deposits, and the amount of ore brought down each year is largely on the increase, as is shown in the following returns from the *Marquette Journal*, of January 16, 1863:

THE IRON PRODUCT OF THE LAKE SUPERIOR—SHIPMENTS OF IRON ORE.

Year.	Jackson Iron Company.	Cleveland Iron Company.	Lake Superior Iron Company.	Total Gross tons.
1855................	.....	1,447	......	1,447
1856................	4,497	7,100	......	11,597
1857................	13,912	12,272	......	26,184
1858................	11,104	19,931	......	31,035
1859................	10,662	30,344	24,668	65,679
1860................	41,286	42,696	33,016	116,998
1861................	12,919	7,311	25,200	45,430
1862................	42,767	35,244	37,710	115,721
Total amount shipped to date........................				414,091

These ores are the peroxide, or specular variety, often nearly chemically pure, but generally contain a small quantity of silicious matter. There is hardly a trace of sulphur, phosphorus, or titanic acid, and the product is a fine, tough, fibrous iron. No mining is required, for the ores lying in great knobs, or ledges, are worked in an open quarry. These ores are in great demand in western Pennsylvania and northern Ohio, where they are mixed with the carbonates of the Coal Measures, by which combination the quality of the iron is vastly improved.

There is no portion of the North-West which will be more benefited by an ample water-communication, than the Iron Region of Lake Superior. With cheap freights, these ores can be sent to the sea-board, or wherever cheap fuel obtains. As they yield over 50 per cent. of pure iron in the working, they will bear a long transportation. A railway is about to be constructed, uniting the head of Bay du Noquet of Lake Michigan, with the mouth of Chocolate river of Lake Superior, the distance being 46⅔ miles. This is an important link in internal communication; first, as affording an additional outlet for these ores; second, as shortening the voyage to Lake Superior five or six days, and avoiding the difficult navigation of the St. Mary's river; third, as protracting the water-communication each season with that region at least six weeks; and fourth, as enabling us to preserve an uninterrupted intercourse with that region, in the event of a war with Great Britain.

SALT-BASIN OF MICHIGAN.

Within the last few years a valuable salt-basin has been developed in the region of Saginaw Valley, in the Lower Peninsula of Michigan, which is estimated to be 17,000 square miles in extent. The product in 1862 had reached 1,270,000 bushels—the result of twenty-two wells—and the number has now reached about one hundred, whose product for the present year is estimated at 4,000,000 bushels, which will find its principal market in the Western States. The product of the Onondaga Salt Springs, which last year reached 9,054,000 bushels, has heretofore been largely absorbed by the North-West, having been used in Nashville and even Leavenworth; while not less than 1,360,000 bushels were shipped to Chicago and Milwaukee.

GOLD DEPOSITS OF THE ROCKY MOUNTAINS.

Recent geological explorations would seem to indicate that the Rocky mountains are auriferous throughout their entire range in the United States, from Mexico on the south, to the British Possessions on the north, extending from latitude 31° 30′ south, to 49° north, and from longitude 102° to the Pacific coast, embracing portions of Dakota, Nebraska, Colorado, all of New Mexico, with Arizona, Utah, Nevada, California, Oregon, and Washington Territories. The region comprises, according to the Commissioner of the General Land Office, seventeen degrees of latitude, or a breadth of 1,100 miles, from north to south, and is of nearly equal longitudinal extension, making an area of more than 1,000,000 square miles. It is traversed from north to south, first on the Pacific side by the Sierra Nevada and the Cascade mountains, then by the Blue and Humboldt mountains, Wasatch, the Wind River chain, and the Sierra Madre, stretching longitudinally and in lateral spurs, crossed and linked together by intervening ridges.

In addition to gold, Nevada and New Mexico are rich in silver.

In the Salmon river district, the yield of gold for the present year is estimated at $20,000,000; while the whole yield of the region is estimated by the Commissioner at $100,000,000.

These figures seem startling, when it is considered that, prior to the discovery of the California mines, the annual gold-product of the world was estimated at only $18,000,000.

This region is rapidly filling up with adventurers, who are to be fed and clothed, and supplied with all the comforts and conveniences of civilized life. They must be bound to the parent States not simply by the ties of early association, but by those of interest. In all mining enterprises, collossal machinery is required; the steam-engine must be employed to pump, to lift, to crush, to wash, and to perform a vast variety of processes which human hands could hardly accomplish. It was politic to extend to this region a Pacific railway; it will be politic to afford to its inhabitants, as far as practicable, a cheap water-communication. It is a matter of deep interest to them whether their supplies, for two-thirds of the distance, are moved by rail or by water.

These are the elements of a commerce, which, although in its infancy, has already assumed gigantic proportions, and is clamoring for additional outlets.

COST OF TRANSPORTATION.

This subject has been elaborately investigated by McAlpine, while State Engineer of New York, with the following results:

	Mills per ton per mile.
Ocean, long voyage	1½
" short "	2 to 6
Lakes, long "	2
" short "	3 to 4
Hudson river	2½
Mississippi and St. Lawrence	3
Erie Canal, enlarged	4
Ordinary canals	5
Railroads, ordinary grades	12½ to 13⅛

Assuming these rates as being substantially correct, it will be seen that the relative cost of transportation by rail, as compared with the other modes of conveyance, is as follows:

	Per Cent. greater.
By Rail, over Ocean Transportation	733.3
" " Great Lakes "	525.0
" " Mississippi and St. Lawrence Transportation	316.6
" " Hudson "	400.0
" " Illinois Improvement "	257.1
" " Erie canal enlarged "	215.0
" " Ordinary canal "	150.0

These are the elements, from which any one interested in this subject, can compute the practical effects upon the productive industry of the country, and the enlarged area it will give to cultivation,—the result of increased avenues of communication between the Mississippi and the sea-board. The producer will have new motives to multiply his crops, while to the consumer will be held out the prospect of cheap bread. Viewed in its true light, the Railroad interest can interpose no valid objection. With industry active and remunerative, travel will increase, as well as the consumption of those articles which require a rapid transit, and for which this mode of conveyance is specially adapted. The resources developed along the lines of communication will more than compensate for any loss of through traffic, and the equilibrium between out-going and returning freights become far more constant than it now is.

STATEMENT SHOWING THE RATES OF TRANSPORTATION BETWEEN THE MISSISSIPPI RIVER AND NEW YORK FOR 1862; ALSO THE COST WITH A COMMODIOUS WATER-COMMUNICATION.

FROM ALTON	TO CHICAGO— Illinois River Improvem't.				TO BUFFALO— By Lake.				TO ALBANY— Erie Canal enlarged.				TO NEW YORK— Hudson River.				TOTAL.		AMOUNT SAVED. 1. Over Summer Rates. 2. Over Winter do.
	Cost per Ton per Mile.	Distance.	Cost per Bushel.	Charged per Bushel.	Cost per Ton per Mile.	Distance.	Cost per Bushel.	Charged per Bushel.	Cost per Ton per Mile.	Distance.	Cost per Bushel.	Charged per Bushel.	Cost per Ton per Mile.	Distance.	Cost per Bushel.	Charged per Bushel.	Cost.	Charged.	
	Mills.	Miles.	Cts.	Cts.	Mills.	Miles.	C	Cts.	Mills.	Miles.	Cts.		Mills.	Miles.	Cts.	Cts.†	Cts.	Cts.	Cts.
Corn	3½	320	3.0	14‡	2	950	5.3	9.6	4	350½	3.9	....	2½	151	1.0	13.3	13.2	36.8	23.6
Wheat	"	"	3.2	15	"	"	5.7	10.6	"	"	4.2	....	"	"	1.0	15.4	14.2	41.0	26.8
Flour—bbls	"	"	11.2	35	"	"	19.0	43.0	"	"	14.0	....	"	"	3.7	70.0	47.9	148.0	100.0
BY RAIL.*	Cts.	Dis.	Cts.	Cts.	Cts.	Dis.	Cts.	Cts.	Cts.	Dis.	Cts.	Cts.	Cts.	Dis.	Cts.	Cts.	Cts.	Cts.	Cts.
Corn	1¼	256	8.9	14	1¼	513	17.9	27.5	1¼	380	13.3	20.0	1¼	144	5.0	7.0	47.6	69.4	56.2
Wheat	"	"	9.6	15	"	"	19.2	29.5	"	"	14.2	21.4	"	"	5.4	8.5	48.4	74.4	60.2
Flour	"	"	32.0	50	"	"	64.1	98⅓	"	"	47.5	71⅓	"	"	18.0	27.0	161.6	246.6	198.7

* For Six Months, during the suspension of Navigation. The cost is given by rail; but, in the last column, from the amount charged is deducted the cost by water.

† Amounts charged between Buffalo and New York included in the same column.

‡ Existing rates by rail.

To illustrate the immense saving to be effected on the cost of transportation, by the opening of these two enlarged avenues between the Mississippi and the sea-board, the Committee have compiled, with great care, the preceding table, which shows the ordinary freights by water, and by rail; and what would be the actual cost, with a commodious water-communication. The result is, as compared with the summer-rates, a saving of one-half; and, as compared with the winter-rates, a saving of two-thirds. These rates amount to a virtual prohibition, in ordinary times, on the shipment of corn, a hundred miles west of Lake Michigan.

It will thus be seen that the actual cost of transporting a bushel of corn from the Mississippi to the Atlantic, would be $13\frac{2}{10}$ cts.
To which add two elevator charges, - - - 1 "
Tolls, say $1\frac{1}{2}$ cts. on each improvement, - - 3 "

$17\frac{2}{10}$ cts.

In the Chicago market in 1861, between June and December—the most active period of navigation—the price of corn vibrated between 20c and 24c. The cost of transportation from the Mississippi to Chicago was 16 cents; while the cost of gathering, shelling, and hauling to a station, would exceed the difference between the rate for transportation and the Chicago price; so that, if a person had been gratuitously offered a given amount of corn, to be gathered west of the Mississippi, on condition that he sent it to the Chicago market, he could not have afforded to accept the gift. That year, the freights paid by one corporation on more than 1,500,000 bushels, were $15\frac{18}{100}$ cents from Chicago to Buffalo, and $17\frac{66}{100}$ from Buffalo to New York, making, in all, $32\frac{84}{100}$ cents a bushel.

The subjoined statement shows the distances from some of the principal commercial points, to the mouth of the Illinois river; also, the cost of transporting a bushel of corn, *via* the improved water-communication:

PLACES.	Distance.	Freight to Illinois river.	Freight to New York.
	MILES.	CENTS.	CENTS.
Ft. Union	1,900	15 8–10	32 6-10
St. Joseph	508	4 2-10	21 4-10
Pittsburgh	1193	10 2-10	27 4-10
St. Paul	634	5 4-10	22 6-10
Davenport	319	2 7-10	19 9-10
New Orleans	1,320	12	29 2-10
Memphis	448	3 8-10	21
St. Louis	50	4-10	17 6-10

From this table it will be seen what an immense scope of country will be made as accessible to New York, as Chicago and Milwaukee are at this time. The pioneer, upon the farthest verge of settlement in the vicinity of a navigable river, will find his crops as remunerative as those of the Illinois farmer a hundred miles from the Lakes.

The construction of these works would add untold millions to the national wealth, and communicate an impetus to agriculture and settlement, such as has not been paralleled even in the past history of this region.

NECESSITY OF ADDITIONAL OUTLETS.

The testimony of commercial men is concurrent, that the existing avenues of communication between the Lakes and the sea-board are inadequate to accommodate the traffic. For the past two years, the warehouses on the Lakes have been, during the active period of navigation, gorged with freight; the rolling-stock of the railways has been worked to its full capacity; every craft that could float upon the Lakes has been put under steam, or canvas; and the locks of the New York canal have proved inadequate to expeditiously pass the throng of boats; so that the voyage which ought to have been performed in nine days has been protracted to fourteen, and even twenty.

THE NEW YORK CANALS.

The Erie canal is the principal outlet through which the cereals of the North-West are conveyed to tide-water. Its dimensions are 70 feet wide and 7 deep, with locks 13 feet wide and 110 feet long, whose contents are about 13,800 cubic feet. The total length is 350.58 miles.

To show the extent to which its transporting capacity is taxed, your Committee beg leave to call attention to some facts contained in the Annual Report for 1862, of the Auditor of the Canal Board of the State of New York.

The total tonnage, its value, and the tolls collected on the canals, during 1862, were as follows:

Tons.	Value.	Tolls.
5,598,785	$203,234,331	$5,188,943.

The value of Western products, passing through the Erie and the Champlain canals to tide-water, has increased more than 100 per cent. within the past four years.

STATEMENT SHOWING THE EASTWARD-BOUND TRAFFIC OF THE ERIE AND CHAMPLAIN CANALS FOR FOUR YEARS ENDING 1862.

YEARS.	1859.	1860.	1861.	1862.
Tons....	2,121,672	2,854,877	2,980,144	3,402,709
Value................	$53,175,312	$78,798,617	$81,332,759	$111,176,568

The proportion between *Way* and *Through* traffic was about 1.8.

Turning to wheat, it will be found that less than one-tenth was local, while more than nine-tenths were drawn from the granaries of the North-West.

STATEMENT SHOWING THE MOVEMENT OF FLOUR THROUGH THE NEW YORK CANALS TO TIDE-WATER FOR FOUR YEARS ENDING 1862.

(*Wheat reduced to Flour.*)

YEAR.	Bbls. West.	Bbls. N. York.	Bbls. arriving at tide-water.
1859	2,210,620	....	1,925,402
1860	4,344,387	737,321	5,081,708
1861	6,712,233	747,822	7,457,225
1862	7,516,397	843,685	8,360,082

The following is a statement of the total receipts of flour and corn—wheat being reduced to flour—at tide-water at New York, for 1862:

	Flour, barrels.	Corn, bushels.
By Canal.......	8,360,082	32,670,000
By Rail	2,617,923	450,000
TOTALS	10,978,005	33,120,000

In 1855, Mr. Jarvis, a distinguished engineer of New York, predicted that, in fifteen years, there would be an eastward movement of five millions of tons, the surplus products of the North-West. His prediction has been verified in seven years, or within one-half the time assigned.

COST OF MOVING THE CROPS.

The amount of eastward-bound tonnage, including flour, conveyed over the three principal trunk lines of railway, in 1862, is shown in the following

STATEMENT FROM OFFICIAL REPORTS.

	Tons Eastward.
New York Central Railroad	616,177
New York and Erie Railroad	471,314
Pennsylvania Railroad	502,884
	1,590,375
To which add by Erie and Champlain Canals	3,402,709
TOTAL	4,993,084

Now, if all of these Western commodities were reduced to as compact a form as flour in barrels, and we were to suppose that thirty per cent., as in the case of the cereals, came from the west of Lake Michigan, thirty per cent. from east of that lake, and forty per cent. from the longitude of Chicago, and that the freights charged were in proportion to those on flour, during the past season, they would amount to more than $56,400,000, as the cost of transferring the annual products of the North-West to the seaboard. To this should be added the freights on about $11,000,000, which found their way through Canada. With improved facilities, such as have been proposed, costing not to exceed $17,000,000, the saving in the movement of a single crop would amount to $30,000,000.

CAPACITY OF EXISTING OUTLETS.

It must be borne in mind that these great thoroughfares are fast approaching their full capacity for transportation. The capacity of the enlarged Erie canal was rated by McAlpine at seven million tons; but this was on the supposition that the enlarged locks would expeditiously pass the boats. Already the tonnage of the main line and its affluents has reached five and one-half millions, and the voyage, which ought to be performed in nine days, is protracted to fourteen, and even twenty.

To show that these views are not exaggerated, your Committee beg leave to refer to the statement of a highly respectable body of gentlemen, representing the Corn Exchange of New York, and the Board of Trade of Buffalo, submitted to the joint Committee on

Canals, of the New York Legislature, in April last, in which it is shown that the capacity of the locks has been reached the past season at 2,900,000 tons, and that there is an improbability of increasing the movement of tonnage by increasing the number of boats.

"The fact was shown that during considerable portions of the past three years, the Erie Canal had been taxed to its utmost capacity, not from deficiency in its main trunk, but from the impossibility of passing more boats through its locks; that while the channel of the canals was sufficient to be navigated by boats of six hundred tons burthen, the present locks could pass boats of about two hundred tons only; that while the channel of the canals in question was 70 by 7, the locks were but 97 feet by 18; that multiplying boats would not increase the transportation of tonnage, for the reason the limit of lockage had been already reached; that while the channel of the Erie and Oswego canals (with resources at command), were probably sufficient for the transportation of twenty millions of tons annually, the capacity of the present locks had been reached the present season at 2,900,000 tons.

"The improbability of increasing the movement of tonnage by increasing the number of boats (the locks remaining as now), was strikingly illustrated by tables furnished by the Auditor of the Canal Department, viz.: The lockages for the three most active months of 1860 (September, October, and November), were 15,420 at Frankfort (near Utica, which locks are double). For the year 1861, there were added 619 new boats (and the fact is notorious that, during such remunerative seasons as 1860, 1861 and 1862, few boats go out of existence, but are repaired and kept in use); yet the lockages for the same months, at the same place, for 1861, were but 15,585, showing an increase of 165. For the year 1862, were added 850 new boats to the number in use in 1861 (an increase of 1,469 over the number of boats in use in 1860), yet the lockages, by the most extraordinary exertions, by employing additional men, stationary power at the locks, abundance of water, and with singular exemption from breaks, were brought up to but 17,083 during the same period—an increase upon 1860 of 1,663; whereas, with adequate locks, the increased number of boats, for 1861 should have shown an increased lockage of 3,714, instead of but 165, and for the year 1862, a lockage of 8,994, instead of but 1,663. It should be borne in mind, while these lockages are actual, that the season of 1860 was cut short of 1861 and 1862 fully two weeks, by early ice; that in 1860, boats which cleared at Buffalo on the 17th of November, were frozen in west of Albany, while in 1861 and 1862, boats reached New York which cleared at Buffalo as late as November 26th; therefore, had canal navigation of 1860 remained uninterrupted by ice as late as was that of 1861 or 1862, there is every probability the lockages of 1860 would have been as great as those of 1861 and 1862. These lockages represent the number of trips made by boats during the three months under examination; therefore, the differences between the number of actual lockages proportional to the number of boats employed, compared with those of 1860, multiplied by the average tonnage of canal boats, unmistakably represent the loss of tonnage to the canals during that period.

"The inadequacy of the locks to the present channel of the canals was further illustrated by the many miles of boats constantly accumulated at Rochester, waiting their turn at the Brighton lock, so called; and at Syracuse, at the first lock east of the junction of the Oswego canal, showing that while these boats had passed readily along the levels, they suffered detention only at the locks; thus, while ten to twelve days should be ample time to run a loaded boat from Buffalo to New York, eighteen to twenty-two are now required, consequently, a loss in time of nearly thirty-three per cent."

MULTIPLICATION OF RAILROADS AN INADEQUATE RELIEF.

It is not to be supposed that the trunk lines of railway can accommodate this growing commerce, for the reasons, first, that it will not bear this expensive transit; and second, that with their large passenger-business, and fast freight-lines, for the conveyance of merchandise and perishable articles, such as fresh meats, vegetables, etc., constituting the daily food of the great cities, they combined could not convey eastward an additional million of tons. The four great American outlets, then, (the New York canals, the New York Central, Erie, and Pennsylvania railways), have a capacity, at the highest estimate, to accommodate only about two and one-half additional millions of tons; and that, too, in view of an eastward-bound commerce through those channels, whose increase will at an early day reach the full limits of their capacity.

These facts demonstrate the absolute necessity of additional outlets—cheap, commodious and expeditious—for Western commodities, or production, up to the point already attained, must cease.

To relieve the existing glut in transportation, it has been proposed to construct additional railways.

When railroads were first introduced, it was supposed by many that they would supersede canals; and that expeditious transport, though at an increased cost, would counterbalance the cheapness of water-communication. Experience, however, has shown that this supposition was fallacious, and the relative advantages of these two modes of transport are now fully understood; and perhaps, there is no more striking example of this than in the State of New York, where the Central railroad and the Erie canal stretch coterminous through that State. The one is employed for the expeditious transportation of passengers, of perishable articles requiring immediate consumption, and of those to which an enhanced value has been communicated by the industry of man;

the other for those crude and bulky articles, which, in the absence of canals, would yield to the producer little or no return. In the one case, time becomes an element of value, for which the consumer is content to pay; in the other, it is a matter of slight importance.

The Committee have already shown that, under the existing rates of transportation, the export of corn is bounded by the western shore of Michigan; but with an improved water-communication, it would be extended to the farthest confines of settlement. Now, if the corn of the Western farmer, when charged with freights, through a water-communication, of twenty-five cents a bushel, between Chicago and New York, yields him no profit; if his wheat, charged also, with twenty-seven cents a bushel, is excluded from the markets of the world, except in case of public famine, what relief is it to him to construct additional modes of conveyance, on which the charges are 100 per cent. higher than by the existing routes?

EASTERN DEPENDENCE ON WESTERN BREAD-STUFFS.

The cereals of the North-West having found their way to New York, it is proper to trace their distribution;—to show how much is required to feed the inhabitants of the sea-board, and how much remains to form the basis of our foreign commerce.

By the census returns of 1860, it appears that New England raises wheat barely sufficient to feed her population three weeks; New York, six months; Pennsylvania is about self-sustaining; while Ohio yields a surplus of less than 3,000,000 bushels. In these States, during the last decade, there was a falling off in the amount of production to the extent of 6,500,000 bushels, while the increase in the North-West exceeded 55,000,000 bushels.

FOREIGN EXPORTATION.

After supplying the deficiencies of the sea-board States, the North-West has a surplus of bread-stuffs and provisions for exportation, which, in four years has increased in value from $38,300,000 to $122,650,000, which is—exclusive of specie—nearly 70 per cent. of our foreign exports. To this may be added $11,100,043, largely grain, which went out through Canada, making the aggregate over $133,750,000. This is shown by the annexed statement taken from the reports of the Secretary of the Treasury.

STATEMENT SHOWING THE VALUE OF THE DOMESTIC EXPORTS OF THE UNITED STATES, EXCLUDING SPECIE, FOR THE LAST FOUR YEARS.

ARTICLES.	YEAR ENDING JUNE 30.			
	1859.	1860.	1861.	1862.
Total Exports..............	$278,392,080	$316,242,423	$204,166,299	$181,875,988
Bread-stuffs and Provisions .	38,305,991	45,271,850	94,866,735	122,650,043
Domestic Exports to Canada (largely grain)	21,769,627	11,264,590	11,016,664	11,100,069
Foreign Exports...........	6,884,547	2,918,524	2,505,735	....

These statistics show that the export of bread-stuffs and provisions in two years increased nearly 180 per cent. in value; and in three years, 220 per cent.

MATERIAL PROSPERITY OF NEW YORK.

This vast mass of vegetable and animal food, moving from the West to the East, with all the regularity of an ocean-current, has enriched the region along its route. It has been the main source of revenue to the New York canals and railways. It is the principal mine, from which New York city has derived her opulence and commercial greatness. The highlands of the Hudson are the gateways of a commerce, such as Venice, in her palmiest days, never dreamed of. She, not simply by her geographical position, but by the extended system of internal improvements, constructed by the State and public corporations, has been enabled to exact tribute upon nearly every article, whether produced or consumed by the North-West; upon the raw material, as well as the manufactured fabric; upon the proceeds of the outward voyage, as well as the return cargo. She has made herself the connecting link between two continents; the centre to which converge all the great lines of trade; the mart to receive and distribute the imports and exports of a continent.

The Committee might go further, and show how much more lucrative to New York has been her commerce in corn, than in cotton; how the one, from the time it started from the banks of the Mississippi, until it arrived at the sea-board, scattered riches in its path; how the other, leaving a Gulf port, simply touched at New York, and then departed for some English mart, leaving behind no substantial benefit. The one is like a noble river, fertilizing the region through which it flows; the other as barren as the ocean on which it floats.

In view of these facts, public sentiment requires, and has a right to demand, that the State of New York shall hold this great thoroughfare—this connecting link between the East and the West—not for local aggrandizement, or State revenue, but as the trustee of the nation; and impose only such tolls on commerce as shall be required to preserve the integrity of the work, and ultimately pay the cost of construction.

FOREIGN DEMAND FOR THE SURPLUS OF THE NORTH-WEST.

The question of demand and supply remains to be considered; whether the European nations will require Western bread-stuffs and provisions only to a limited extent, and that, therefore, production, up to that point, must cease; or whether they will absorb our surplus, however great. In reply, it may be stated as a general truth, that there is not an instance in human history, so closely does population press on the means of sustenance, of a vast accumulation of food, beyond the wants of consumers.

The existing population of the European States is estimated at 280,000,000, of whom 150,000,000 are consumers of cereals to the amount of nearly 1,000,000,000 of bushels. The means to further production are limited by the obstacles interposed by nature, against which it is in vain for man to contend,—inhospitable mountains, barren wastes, and irreclaimable marshes. The most serious obstacle, therefore, to the increase of population will be the limits placed on the production of human food; but, through the equalizing effects of commerce, it is safe to presume that there will, at all times, exist an active demand for our surplus bread-stuffs, in exchange for the peculiar products of their soil, climate, and industry; and this demand will keep pace with the density of their population. Speculations, therefore, as to the probability of glutting the foreign market seem idle and misplaced.

The dependence of Great Britain upon foreign supplies each year becomes more apparent. In 1855, it amounted to 59.02 per cent.; and in 1860, it rose to 88 per cent.

STATEMENT, FROM OFFICIAL SOURCES, SHOWING THE AMOUNT OF WHEAT AND OTHER GRAIN, AND FLOUR, IMPORTED FROM THE UNITED STATES AND OTHER COUNTRIES INTO THE UNITED KINGDOM OF GREAT BRITAIN FOR FOUR YEARS ENDING 1861.

	1858.	1859.	1860.	1861.
WHEAT.				
	Qrs.	Qrs.	Qrs.	Qrs.
United States..............	594,644	36,906	1,499,385	2,507,744
Other Countries............	3,647,075	3,964,016	4,381,573	4,405,071
Total..................	4,241,719	4,000,922	5,880,958	6,912,815
GRAIN—OTHER KINDS.				
	Qrs.	Qrs.	Qrs.	Qrs.
United States.	399,807	9,948	475,178	1,779,652
Other Countries............	5,545,739	5,307,813	6,649,484	5,586,587
Total..................	5,945,546	5,317,761	7,125,662	7,366,239
GRAIN—ALL KINDS.				
	Cwt.	Cwt.	Cwt.	Cwt.
United States..............	1,764,795	216,462	2,254,238	3,795,865
Other Countries............	2,091,332	3,111,862	2,831,988	2,358,073
Total Cwt.	3,856,127	3,328,324	5,086,220	6,152,938
In Quarters............	1,101,750	950,949	1,453,205	1,757,982
FLOUR.				
	Qrs.	Qrs.	Qrs.	Qrs.
United States..............	1,098,871	98,752	2,143,451	3,591,991
Other Countries............	4,244,598	4,853,119	5,190,713	5,078,806
Total..................	5,343,469	4,951,871	7,334,164	8,670,797
GRAIN AND MEAL.				
	Qrs.	Qrs.	Qrs.	Qrs.
United States..............	1,500,481	109,275	2,624,005	5,398,176
Other Countries............	9,793,224	10,161,499	11,873,971	10,696,738
GRAND TOTAL..........	11,293,705	10,270,774	14,497,976	16,094,914

The imports for 1862 were, according to the *London Gazette*, as follows:

Wheat, Quarters	9,542,359
Indian Corn, Quarters	2,751,261
Flour, Cwt	7,314,331

CONTINENTAL SOURCES OF SUPPLY.

The great European port for wheat-shipping is Dantzic, on the Baltic. The wheat is raised in Galicia and Poland, from five to seven hundred miles inland, and brought to the sea-board in flat-bottomed boats, suited to the navigation of rivers usually shoal, and abounding in rapids,—a mode of conveyance both tedious and expensive, costing from 6*s.* 6*d.* ($1.56), to 9*s.* 2*d.* ($2.20), per quarter to place it at Dantzic. These rates would be from 16 cents to 23 cents a bushel;—higher than the rates between the Mississippi and the Atlantic, with an improved navigation.

Another great source of supply is the Black Sea ports. The Dneister, the Dneiper, the Don and Volga are navigable, but abound in shoals and rapids. Wheat is sent to Odessa and Kertch by these streams; and by land, it is brought to market in waggons, often from a distance of many hundred miles.

The route between Odessa and Liverpool is circuitous, and consumes as much time as is required to cross the Atlantic. It is necessary that the voyage be performed in the winter season, in consequence of the heated waters of the Mediterranean; for it often happens that cargoes of wheat arriving in summer have to be removed with the pick-axe. The price on board at Odessa considerably exceeds 40*s.* per quarter, and the expense of importing is from 16*s.* to 18*s.*

In 1861, England imported grain of all kinds—

From Southern Russia	1,282,127	qrs.
From the Baltic	1,532,783	"
From British America, a considerable portion of which was derived from the United States	1,188,839	"
From the United States	5,398,176	"

From no other country did she derive a million quarters.

ABILITY TO COMPETE WITH FOREIGN MARKETS.

The average English price of wheat for the last quarter of a century, has been 54*s.* 6*d.* per imperial quarter of 70 lbs., which would be equivalent to $1.42 for an American bushel. The continental price is 6*s.* 6*d.* less, that being the cost of transportation to England, per quarter, which, on an American bushel, would equal $16\frac{7}{10}$ cents.

Now, with an improved water-communication, the cost of shipping a bushel of wheat from the Mississippi to New York, and thence to Liverpool, would be—

	Cents.
By Canals, Lakes and Rivers	14.2
Tolls, say	3.0
Three Elevator-charges	1.5
Insurance and Commissions	1.7
Ocean, 3,150 miles, at $5 per ton	15.0
Cost to Liverpool	35.4
Where it would be worth	142.0
Netting to the Shipper	106.6
To which add Premium on Exchange	10.7
Making	$1.17.3

Which would be a discrimination of only 8 cents against the American producer, as compared with the continental price, and would make the average price of wheat $1.06 on the banks of the Mississippi; $1.00 at St. Paul; and 80 cents at Fort Union, with the allowance of a margin of 10 per cent. for contingencies.

When we consider the character of the wheat-growing region of the North-West, the cheapness of the lands, the fertility of the soil, and the facility with which it is cultivated,—that all of the processes of sowing, reaping, harvesting, binding, threshing and winnowing, are done expeditiously by machinery, the American farmer may successfully enter the lists of European competition, and contend for a monopoly of the provision-market of the world. He need not depend on any accidental deficiency in the crops of Europe, but rely on a nearly unvarying market for all his surplus crops abroad. This traffic in human food will prove a greater power than ever cotton was, and give us a strength, and position among the nations of the earth, far above what we have already attained.

The Committee have thus imperfectly sketched the great features of this commerce; but, in traversing a field so vast, have been compelled to leave out many subordinate details. The facts adduced show how essential the construction of these works is to the future prosperity of the North-West, and to the whole country.

IMPORTANCE OF A SHIP-CANAL.

The Committee have thus adverted to the magnitude of the products of the North-West, the burdens to which they are subjected in their passage to the sea-board, and the extent to which they enter into our external commerce, and contribute to the national wealth. While almost every other industrial interest of the country,—the coal and iron of Pennsylvania, the manufacturing of New England, and the salt of New York,—is protected by discriminating duties of thirty per cent., we search the statute-book in vain for any legislation, which tends directly in aid of agriculture—the main-spring of all our prosperity.

It is proposed to consider this question in three aspects, viz. :

I. NATIONAL,

As tending to bind together different parts of the Union, and uniting the people by the ties of mutual interests and social connections ; and as developing the resources of distant regions, and thereby contributing to the national greatness.

II. COMMERCIAL,

As affording a cheap and expeditious communication between regions widely separated, and as admitting of a free interchange of the products of different climes, and of different industries, giving activity to labor, and a profitable return to capital.

III. MILITARY,

As connected with the defense of the country, using such a communication as a means of transferring gun-boats from one system of waters to another, and of rapidly concentrating them at points widely asunder, thus making a small armament as efficient as a large one.

NATIONAL ASPECTS.

CONSTITUTIONAL POWER OF CONGRESS.

The Constitution empowers Congress to do all necessary acts to provide for the COMMON DEFENSE, and to promote the GENERAL WELFARE.

Mr. Jefferson, in 1801, on assuming the duties of the Presidency, announced as among the leading objects of the Constitution—" the encouragement of agriculture, and of commerce, its handmaid."

Mr. Madison, the Father of the Constitution, in 1809, when called to the same exalted position, uttered a similar declaration,—" to promote by authorized means, improvements friendly to agriculture, to manufactures, and to *external* and *internal* commerce;" and, in 1816, he called the attention of Congress to the importance of devising a comprehensive scheme of roads and canals, " such as shall have the effect of drawing more closely together, every part in the common stock of national prosperity."

As far back as 1807, Albert Gallatin, one of the most far-seeing and sagacious of our statesmen, as Secretary of the Treasury, submitted an elaborate report to the Senate on the importance of constructing roads and canals by the Government, as a means of affording speedy and easy communication between remote parts of the country, to facilitate commercial and personal intercourse, and to unite the people by a still more intimate community of interests. In that report he states,—"No other single power of government can more effectually tend to strengthen and perpetuate that union, which secures external independence, domestic peace, and internal liberty."

There has, from the adoption of the Constitution, existed a class of men who viewed with extreme distrust every exercise of power on the part of the Government, for the promotion of the general welfare. All schemes to facilitate communication between remote territories, to remove obstructions in the pathway of commerce, or to develop particular branches of industry, have been pronounced unconstitutional; while, on the other hand, all attempts to fortify the the approaches to our territory, to build up and equip an efficient navy, and to maintain a well-disciplined army, have been denounced as a wasteful expenditure of money. But the events of the last two years have taught us a far different lesson; and that ours is not an exception to the history of other nations who have preserved their integrity only by the strong arm of power.

"To govern an extended empire," using the words of Gibbon, with a slight alteration, "requires a refined system of policy; in the centre, a strong power, prompt in action and rich in resources; a swift and easy communication with the extreme parts; fortifications to check the first effort of rebellion; a regular administration to protect and punish; and a well-disciplined army to inspire fear, without producing discontent and despair."

THIS POWER REPEATEDLY EXERCISED.

This power has been repeatedly exercised by Congress;—for example, in the construction of the National Road, which was the first commodious channel of communication between the Valley of the Mississippi and the sea-board; in subscriptions to various canals,—the Louisville and Portland, the Delaware and Chesapeake, the Potomac and Ohio, and the Dismal Swamp canals; and more recently in the munificent grant of bonds and lands in aid of the construction of the Pacific railroad;—a measure called for by every consideration of national unity, internal commerce, and military defense.

NATIONALITY OF THIS COMMERCE.

The commerce which floats upon a river like the Mississippi, draining half a continent; or upon the Great Lakes, whose shore-lines are longer than those of the sea-board States; or is poured through an artificial channel like the New York canal, is as much national as that which is wafted over the Atlantic. When it is shown that eight-ninths of the cereals are derived, not from a single State, but from a group of States; and are moving, not to a local market, but to the markets of the world; furnishing to the navigating interest the outward-bound freight as well as the return cargo, and conferring a direct benefit on the national finances; and when the proceeds of these products are traced through all the ramifications of trade, it is evident that it is not simply the citizen of one State, but the Western producer, the consumer at home and abroad, the navigator, the importer, the consumer of foreign fabrics, and the Government itself, all have a direct interest in the result.

It is a measure whose benefits are not to be circumscribed by State lines, but one which connects three distinct systems of navigation, and renders them available for external and internal commerce, for national unity, and military defense.

Every one is aware how largely the topographical features of a country influence its inhabitants in their social habits, their modes of thought, and business pursuits. The words of Cowper contain a fund of political philosophy:

> "Lands intersected by a narrow frith
> Abhor each other. Mountains interposed
> Make enemies of nations, who had else,
> Like kindred drops, been mingled into one."

But bridge these friths, tunnel these mountains, making them the great highways of commerce, and you unite the people by the ties of a common interest, which they will consent to sever only under the most pressing necessity.

HOW TO CONDUCT A LONG WAR.

The expenses of the Great Rebellion, reaching not less than $500,000,000 a year, must be levied on our national resources. It is the price which must be paid for the preservation of our national unity. Taxes are assessed upon almost every article that contributes to the wants, or the conveniences of the people; but, however multifarious the tax-schedule, the revenue is derived from two sources alone;—the wages of labor, and the interest of capital. The ability of the people to sustain taxation depends on the reward given to labor, and the profitable employment of capital. The fact, therefore, that the national resources are taxed to so great an extent, for the purpose of sustaining the war, so far from being an argument against appropriations for the objects of internal commerce, is the strongest reason why they should be made. The war must be made to sustain the war.

The most hopeful feature in this contest is the general prosperity of the North, and the general paralysis of the South. In the one section, commerce is active, labor in demand, and wages are almost unexampled; property is constantly advancing in value, immigration has not fallen off, population is increasing, while the operatives in every branch of industry—agriculture, manufactures and mining—are unceasingly occupied. On the other hand, the South exhibits a forlorn aspect. Over wide expanses of territory desolation reigns supreme. With the mouth of the great mercantile river, the Mississippi, in the possession of the Government, and their principal ports blockaded, they are thus cut off from the markets of the world. Their cotton plantations have almost ceased to be cultivated; and while the demand for that staple abroad is unprecedented, at home it is almost valueless. A famine threatens the land; and tumultuous crowds of women parade the streets of their capital crying aloud for bread. Villages are depopulated, refugees flock to the lines of the army demanding protection, and food commands almost fabulous prices.

From the example of Netherlands, in her terrible but successful struggle for nationality, the North can derive a salutary lesson. Although, according to the historian Motley, a war had been raging for a quarter of a century without any interruption, population increased, property rapidly advanced in value, and labor was in active demand. Famine was impossible to a State which commanded the ocean. No corn grew in Holland and Zeeland, but their ports were the granary of the world; and in one month eight-hundred vessels left their havens for Eastern ports alone. While the seaports rapidly increased in importance, the interior towns advanced as steadily. The woolen manufacture, the tapestry, the embroideries of Gelderland, and Friesland, and Overyssel, became as famous as had been those of Tournay, Ypres, Brussels, and Valenciennes. The emigration from other countries was very great; it was difficult to obtain lodgings in the principal cities; new houses, new streets, new towns, rose every day; and when the English embassadors arrived in the Provinces, they were singularly impressed by the opulence and magnificence which surrounded them. The single province of Holland furnished regularly for war expenses alone, 2,000,000 florins a year, besides other extraordinary grants, which seemed only to make it more elastic. A contemporary remarked that "coming generations may see the fortifications erected at that epoch in the cities, the costly and magnificent havens, the docks, the great extension of the cities; *for truly the war has become a great benediction to the inhabitants.*"

By the cultivation of such arts,—domestic industry and external commerce,—they were enabled to carry on a war for eighty years, and bring it to a triumphant issue.

In the midst of a desolating war, Louis XIV completed the canal at Languedoc, connecting the Mediterranean with the Atlantic, which reflected more glory on his reign than all his military conquests.

Napoleon, while combatting with all Europe, devised and executed schemes of national importance, which conferred imperishable benefits on France, and which went far to efface the effects of the ravages of war.

COMMERCIAL ASPECTS.

There is no measure which would so materially benefit our external and internal commerce, as the creation of a ship-canal between the three great systems of navigation in North America,—the Mississippi, the St. Lawrence, and the Atlantic. The Illinois valley, with a summit-level of only eight feet, and with Lake Michigan as an unfailing reservoir, affords an entirely feasible and practicable route; and besides, what is a remarkable fact in the physical geography of the region, its mouth is about the central point of convergence of the three great basins of the Upper Mississippi, with a drainage area of 1,244,000 square miles,—the heart of a great continental system, of which the navigable and unnavigable rivers are the arteries and veins.

Another striking geographical fact is that, taking Memphis and Liverpool as initial points, this route is found to be in a nearly direct line along the great circle of the earth; and is, therefore, the one in which the products of the Great Valley would naturally move to the markets of the world. The New York canal is constructed through a natural depression of the Alleghanies, the most feasible to be found throughout their range from Canada to Alabama. The North-West dates its prosperity from the time of the construction of this work, and its enlargement would form an epoch in a new career of prosperity, compared with which, the past would sink into insignificance.

The facts which have been collated by the Committee show that the products of the North-West feed to a large extent the inhabitants of the sea-board States, and at the same time furnish the bulk of cargoes to our commercial marine; that, exclusive of specie, they constitute in value about 70 per cent. of all of our domestic exports, and in that proportion, contribute to the customs-revenue, in duty-paying articles for which they are exchanged; that, while under the existing tariff, almost every other branch of industry is protected to the extent of 30 per cent., there has been no legislation in aid of bread-stuffs and provisions; that the connecting of these three systems of navigation, under the constitutional power of Congress, by a ship-canal, while its cost would not exceed $17,000,000, would result in a saving of $30,000,000 in the movement of the yearly crops, to be shared alike by the producer and consumer; that its practical effect would be to bring the lands on the outer verge of settlement 2,000 miles nearer the sea-board for all the purposes of market; that products, like corn, now almost worthless for exportation, would be in active demand; that under such a stimulus, the value of the public domain would be greatly enhanced, immigration become active, settlement extended, and our foreign commerce be swollen to an unprecedented extent; and finally, that it would illustrate the great historical truth, that the only method of carrying on a protracted war is to increase the productive industry of the nation.

MILITARY ASPECTS.

Great Britain occupies the northern portion of the continent, with a territory coterminous with our own, stretching from the Atlantic to the Pacific. She has constructed a series of short canals around the rapids of the St. Lawrence, with locks 45 feet wide and 200 feet long, and 8 feet deep; and has connected lakes Erie and Ontario by the Welland canal, with locks 26 feet wide, 150 long, and 11 feet deep, and capable of ready enlargement. She has, also, constructed the Rideau canal, professedly as a military work, by an interior route, between Montreal and Kingston, with locks 33 feet wide and 142 long; and although the channel is only 5½ feet deep, yet it is capable of passing a dangerous vessel, when buoyed up by lighters.

She has a formidable fortress and depot of military and naval stores at Kingston, on Lake Ontario; another at Malden, at the mouth of the Detroit river; and a third at Penetanguishene on Georgian Bay; besides forts more or less impregnable at Toronto, Niagara, Port Stanley, Windsor, and Port Sarnia. Most of these points are intersected by railways, by which a large force can be rapidly concentrated.

To oppose these formidable preparations, we have a few dismantled forts, which a half-hour's cannonading with improved ordnance would batter down, and which, from their weakness, would invite, rather than deter attack; no lines of water-communication by which a war-vessel, larger than a canal-boat can be thrown into the Lakes; no naval or military depots; nothing but a single steamer of 100 tons burden, mounting a single 18-pounder for aggressive or defensive purposes; nor can the number, under the treaty stipulations of 1817, be increased beyond one more on the Upper Lakes, one on Lake Ontario, and one on Lake Champlain.

The richly-laden fleets, bearing the commerce of half the country sea-ward, and the populous cities and towns along the borders of the lakes, are at the mercy of the invader. Not a gun is mounted for their protection, nor is there a harbor of refuge four miles inland.

It may be a startling fact, but it is nevertheless true, that a single battery planted on the Virginian side of the Ohio river below Pittsburgh, and a single gun-boat anchored near the southern shore of Lake Erie, have the power to sever the great arteries of communication between the East and the West. A slight interruption of the calm regularity in the flow of bread-stuffs eastward, would cause a consternation as great as any disaster to our arms. We may repose in fancied security; but should such a disaster occur, succeeding generations would not fail to brand as imbecile, the statesmen of this day, who neglected to defend the nation in its most vulnerable point.

POSSIBILITY OF A RUPTURE.

A year has scarcely elapsed since England, in contemplation of the possibility of a rupture with the United States, began to throw troops and munitions of war into the principal stragetic points; an extended system of fortifications was projected, the local militia was enrolled and equipped—the whole placed under the command of one of the heroes of the Crimean war,—and the aristocratic organ of the nation, the *London Times*, declared that, with the opening of the navigation of the St. Lawrence, England would throw into the Lakes such a fleet of gun-boats as would give her the command of those waters.

It will be recollected, too, that Mr. Seward, as Secretary of State, addressed a letter to the Governors of the States bordering on the Lakes, calling upon them, in the interim of Congress, to take steps to fortify the principal points of approach. It is thus apparent, that both governments regarded a rupture as imminent, and took steps to prepare for it. It is equally evident, that both regarded the Lake-frontier as the theatre of military operations, and commenced a concentration of the *materiel* of war at the principal stragetic points.

While the happening of such an event is greatly to be deplored, still it must be confessed that there are yet irritating questions, which may require to be settled by the arbitrament of war. The fitting out in her ports of vessels to prey upon our commerce, with the tacit knowledge and assent of that Government, cannot but be regarded by us, as it has been by her, a violation of public faith and international comity.

As if in anticipation of hostilities, we again hear of a large force being thrown into Canada, and of the shipment of military and naval stores.

LAKE DEFENSES.

The question recurs, what would be the cheapest and most effectual method of defending the Lakes, and enabling us to assert our supremacy over them in case of war?

The introduction of iron-clad vessels has effected a revolution in naval warfare, and no maritime nation would at this day confine its defenses to stationary fortifications.

The existing forts on the American side, even if furnished with the most approved guns, would probably prove ineffectual to prevent the passage of iron-clads; and besides, stationary fortifications are unfitted for aggressive purposes.

Of the lake-craft, many might be extemporized into war-vessels; but the bulk of them when covered with armor and laden with stores, would be incapable of entering the harbors.

The question of lake-defenses was referred to the Naval Committee of the Thirty-Seventh Congress, who, through Mr. F. A. Conklin, submitted a report which appears to have been written in ignorance of the great hydrographical features of the Lakes, and contains recommendations utterly impracticable.

With harbors along the Lakes admitting vessels of but twelve-feet draft, a fact which seems to have been overlooked by the Committee, they gravely state:

> "Vessels of such a class as could traverse the enlarged canals, would be unequal to a contest with the Roanoke of our own navy, and still more with the La Gloire of the French, or the Azincourt or the Minotaur of the British navy. They ought not to be built for ocean warfare, nor for warfare on the lakes, unless the Government shall be constrained, when the occasion arises, to adopt the lock of a canal as the standard of a man-of-war, and to gauge a contest with England accordingly."

Such a recommendation hardly deserves a passing comment, when it is stated that a vessel of the draft of any of those enumerated—twenty-five feet or more—would be excluded from every lake-harbor, and would be incapable of passing through the straits connecting the respective lakes. If such a policy were adopted, each lake would require its separate fleet, which would be incapable of co-operation.

But, it has been said that the defense of the Lakes is to be made at the mouth of the St. Lawrence. This plan may be acceptable to those residing on the sea-board, and who have no immediate interest in the result; but to those occupying the cities upon the shores, and owning the commerce which floats upon the waters, of the Great Lakes, this plan is far from satisfactory; they will hardly rest secure in trusting to a defense to be made at a distance of two thousand miles. The burning of Buffalo and Black Rock has not yet faded from the recollection of our oldest inhabitants. They have the right, by reason of numbers and the magnitude of the interests involved, to require such an armament as shall enable the Government, at once, in the event of war, to assert and maintain its supremacy on the Lakes. The representatives of the North-West in Congress have at all times cheerfully voted appropriations for fortifications, for ships, arsenals and naval depots, to protect *Ocean* commerce; and now they have a right to demand, as a matter of justice and reciprocal good feeling, that appropriations shall be made for *Lake* commerce,—a commerce wafted on waters whose shore-lines far exceed those of the Atlantic, and whose value far exceeds that of the external commerce of the country. They fail to perceive why one is sectional and the other national,—why one shore, laved by salt water, should bristle with masts and be dotted with forts; and the other, laved by fresh water, should be left defenseless. The West has reason to believe that, when this question is presented, in all of its proportions, the East will return a cordial and emphatic response.

The Congress, up to July, 1861, had appropriated for defense against external aggression, more than nineteen and one-half millions of dollars to the New England States; and more than twenty-nine and one-half millions to the loyal Middle States; while the amount appropriated to the Food-producing States reached a little more than six millions. In the first session of the Thirty-Seventh Congress, 1862, the appropriations for forts, ships, etc., reached fifty millions, not one million of which was given to the North-West.

For the defense of the Lakes is required an iron-clad fleet to co-operate with stationary fortifications. In what manner shall they be introduced? The Naval Committee, with Mr. Conklin as their exponent, have suggested two plans:

1. By constructing a navy-yard on the borders of some one of the inland sheets of water tributary to the lakes.

2. By constructing vessels in parts, and transporting them to the places required ready to be set up.

With regard to the first proposition it may be stated, that, while the Naval Committee admit that it would be an infringement of the treaty-stipulations of 1817 to construct war-vessels on the margins of the Great Lakes, it is difficult to comprehend by what process of reasoning it becomes no infringement to construct them on an inland sheet of water, directly communicating with the Lakes, to be sent down whenever their services shall be required. We apprehend that the British Minister Resident would remonstrate with the Secretary of State, long before the first keel was laid. On the other hand, in enlarging these canals, we are but exercising a right which has been freely conceded to Great Britain.

With regard to the second proposition, it may be stated that, to prepare the materials for constructing two distinct fleets, one for the Upper Lakes and one for the Lower, to be put up whenever the necessity may arise, and to transport them to these waters from parts far remote, would be far more expensive than the cost of opening these lines of internal communication; and besides, the usefulness of these fleets would be restricted solely to these waters.

A third plan has been proposed, and that is to make use of the enlarged canals to transfer our iron-clads from one system of navigation to another, and thus save the expense of maintaining distinct sets of fleets. The Naval Committee, through Mr. F. A. Conklin, maintain that, so far as relates to the enlarged New York canal, it is impracticable. We deem the testimony of ERICSSON, the conceptions of whose inventive skill have saved us from national humiliation, and whose fame will live through the ages with undimmed lustre, of far more importance than any crudely-expressed opinions of this Committee.

"AN IMPREGNABLE WAR-VESSEL OF 25 FEET WIDE, AND 200 LONG, WITH A SHOT-PROOF TURRET, CARRYING A GUN OF 15-INCH CALIBRE, WITH A BALL OF 450 POUNDS, AND CAPABLE OF DESTROYING ANY HOSTILE VESSEL THAT CAN BE PUT ON THE LAKES, WILL DRAW, WITHOUT AMMUNITION, COAL, OR STORES, BUT 6 FEET 6 INCHES OF WATER; AND CONSEQUENTLY, WILL NEED ONLY A CANAL WIDE AND DEEP ENOUGH TO FLOAT A VESSEL OF THOSE DIMENSIONS, WITH LOCKS OF SUFFICIENT SIZE TO PASS IT."

The cost of these enlarged communications, according to the estimates of engineers of the highest capacity and integrity, will not exceed $17,000,000; and yet the Naval Committee, through Mr. F. A. Conklin, without furnishing the country with one iota of proof to impeach the correctness of these estimates, gravely assert that the cost will exceed $45,000,000. With a like facility of pen, these estimates might have been swollen to $100,000,000, if thereby a purpose were to be subserved.

The effects of this rebellion will survive for a generation; and to insure the regular administration of the laws over a portion of the country, will require the maintenance of a force sufficient to put down every display of insubordination. It will be necessary for the Government to control all of the great lines of communication. For this purpose, no means would be so effectual as a class of iron-clad gun-boats drawing from 6 feet to 12 feet of water, and capable of navigating our rivers and entering our harbors. A class of the draft last named, by the aid of lighters, could pass through the Illinois and Michigan canal, from the Mississippi to the Lakes, and *vice versa*, and thus be made available, either to suppress insurrection, or repel invasion.

It may be said that fleets adapted to river-navigation are not adapted to lake-navigation; to this it may be replied that they are well adapted to the defense of the straits, which are the most important lines to be guarded. There are, according to the statement of Admiral Porter, not less than 60 vessels in the United States navy capable of passing the proposed locks of the Illinois and Michigan canal, and others are building of like capacity.

Your Committee, therefore, are of the opinion that the cheapest and most effectual method of lake-defense is, not by the establishment of naval depots, or the building of fleets on these waters, both of which would be construed as a violation of the treaty-stipulations of 1817; nor by the erection of an extended system of land fortifications; but by opening such a line of internal communication that gun-boats may readily be passed from one system of navigation to another, and be made available for defense, alike in the harbors of the Atlantic, on the Lakes, and on the navigable waters of the Mississippi. With these two links in the chain of communication completed, a vessel could be passed, by an internal route, from New Orleans to Chicago, Buffalo, New York, Trenton, Philadelphia, Baltimore, Annapolis, Washington, Norfolk, Richmond, Newbern, and Beaufort, making a distance of 4,300 miles; besides rendering accessible the whole navigable system of the Mississippi and the Lakes. It would, therefore, become a matter of little moment, whether a vessel were built at Brooklyn, Annapolis, Washington, or Philadelphia; or at Pittsburgh, Cincinnati, or St. Louis; the mechanical skill of every section of the country could be called into requisition, and the vessel completed, with little inconvenience, be transferred to the most distant waters.

WAYS AND MEANS FOR CONSTRUCTION.

The bill introduced into the last Congress proposed, for the construction of the Illinois communication, the appropriation of the bonds of the Government to the extent of about thirteen and one-half millions of dollars, redeemable in twenty years, and bearing six per cent. interest per annum, with the pledge of the tolls for the payment of accruing interest, and the ultimate payment of principal, of which the traffic would afford an ample guaranty. The issue of these bonds, thus secured, would subserve all the purposes of a direct appropriation, and would command the confidence of capitalists at a time, when more than ever before there was redundant capital seeking investment. It would not involve the necessity of raising a dollar by taxation.

If it be asked, why does not the State of Illinois execute the work, or confide its execution to a chartered company; it may be said in reply, that the State cannot enter upon the work without first changing her organic law, to accomplish which would require two or three years; and while she is agreed on the policy of surrendering this route to the General Government, to be used as a national highway, it is doubtful whether a like unanimity would prevail with regard to the State taking such action, even if constitutional impediments were not in the way. As to the second inquiry, the State, through her Constitutional Convention, has indicated her policy, in no event to surrender this work to a chartered company.

If it be said that, however meritorious this work, the Government is not in a condition to incur fresh obligations, it may be replied that no debt is formidable, based on a great improvement, whose revenues are ample to meet the accruing interest, and at the same time to create a sinking-fund for the ultimate extinguishment of the principal. The railway debt of the United States exceeds eleven-hundred millions of dollars; and yet the only inquiry of the capitalist, dealing in this class of securities, is, what will be the net earnings? The consolidated debt of Great Britain is so enormous that it will never be paid; yet, based as it is on the opulence of the Empire, it is regarded, the world over, as the emblem of financial stability.

So far as relates to the New York portion of the enterprise, it may be stated that the Legislature of that State, by an act passed April 22, 1862, authorized the enlargement of one tier of locks on the Erie and Oswego canals, provided the expense thereof was paid by the United States; in consideration whereof the last named party should have the perpetual right of passage through said canals, "free from toll, or charge, for its vessels of war, boats, gun-boats, transports, troops, supplies, or munitions of war."

In conclusion, your Committee would state, that this is an enterprise which, in whatever light it is viewed, ought to commend itself to the favorable consideration of the country. In its lowest sense, as a mere pecuniary investment, the bonds of the United States, based on the tolls of the canal, would command the confidence of capitalists. As a commercial scheme, it would enhance the value of the public lands, and communicate a stimulus to agriculture, which would be felt to the farthest verge of cultivation. It would cheapen the price of our daily food, and swell to a vast extent our foreign commerce. As a national measure, it would establish, between the East and the West, closer commercial and political affiliations, and forge a chain which no convulsion could sever; while as a military system, it would prove the cheapest mode of fortifying a long line of frontier, and of controlling an immense inland navigation. In no other way, in the opinion of the Committee, can Congress so effectually, in the language of the Constitution, "PROVIDE FOR THE COMMON DEFENSE," or "PROMOTE THE GENERAL WELFARE."

J. W. FOSTER,
CHAIRMAN.
GEO. F. RUMSEY,
CHARLES WALKER,
WM. McKINDLEY,
R. McCHESNEY,
WM. BROSS,
JOHN B. PRESTON,
Committee.

[B.]

REPORT OF MESSRS. GOODING AND PRESTON,

As to the Plan and Cost of the Improvement of the Illinois and Michigan Canal, and the Illinois River.

Canal Office,
Lockport, Illinois, *May* 30, 1863.

To the Committee on Statistics for the City of Chicago :

Gentlemen :

In presenting our estimate of the cost of enlarging such portions of the present Illinois and Michigan canal, as it is proposed to occupy, and the improvement of the Des-Plaines and Illinois rivers, we shall say nothing in regard to the advantages to result from the proposed work. These have been so clearly demonstrated by others much more able than ourselves, that none except those who are unwilling to believe, can have any doubt upon the subject. But of the feasibility of its construction and its probable cost, many well-disposed persons still have their doubts, and we will, therefore, briefly state a few facts in relation to the character of the work, and our estimate of its cost.

In 1836, the Legislature of this State authorized the construction of the Illinois and Michigan canal, "to be supplied with water from Lake Michigan and other sources," and one of us was appointed Chief Engineer of the work, and acted in that capacity under the Canal Commissioners and Canal Trustees, until its completion in 1848. The other as Assistant and Resident Engineer was actively engaged upon the work, most of the time from 1837 until its completion. We mention these facts to show that we have at least had sufficient opportunities to form correct opinions of the probable cost of the contemplated improvement, when the present canal is to occupy a part of it.

From the Chicago river to the lower dam at Joliet, a distance of 33½ miles, it is proposed to enlarge the present canal to 160 feet in width at the surface, and deepen it upon the Summit to the original level adopted by the Canal Commissioners in 1836, and upon which a large amount of work was executed. This, with a slight additional amount of excavation in the bottom for the first ten miles out of Chicago, will enable boats of full six-feet draft to navigate the canal at the minimum stage of Lake Michigan, from which the supply of water will be directly drawn.

The excavations on the old canal to the requisite depth for the enlarged one, have enabled us to understand, perfectly, the character of the material to be excavated to form the proposed enlargement. The cost of executing the various kinds of work done for a long series of years, and under varying circumstances, we regard as a very valuable guide in fixing the present prices. In fact, such valuable data are seldom or never attainable in estimating a new work, and we have the fullest confidence that our prices are ample, and will afford a fair margin for profits, *provided*, that the average wages of common laborers shall not exceed one dollar per day, the cost of supplies what they have usually been in prosperous times, and that full three years time be given for the completion of the whole work.

The canal from Chicago to Joliet, for which distance the present canal will form a part of the enlarged one, will be much the most expensive part of the proposed improvement, embracing, as it does, the whole of the Summit division.

Below Joliet, the present canal, except for some five or six miles between Marseilles and Ottawa, would be abandoned, and the Des-Plaines and Illinois rivers improved for navigation by locks and dams.

When we furnished the estimate to Gen. Webster, which he indorsed and presented in his report to the War Department, no survey had been made of this part of the line. We had, however, been well acquainted with the Des-Plaines and Illinois rivers for a quarter of a century, knew the amount of lockage, and where the materials must be obtained to construct the dams and locks, etc., etc., and ventured to make an approximate estimate. Many persons opposed to the improvement attacked our estimate, and we were compelled to admit that it had been made without any very precise data. Since then, a survey has been made under our supervision by A. J. Mathewson, Esq., an experienced engineer, whose map and profile of the route from Chicago to La Salle, we submit herewith. We are now able to present an estimate of all parts of the route from Lake Michigan to the Mississippi river, which we trust will be satisfactory to the friends of the proposed improvement. Its enemies will not, probably, be satisfied with the facts in the case, for they prove incontestably that no improvement of anything like equal importance, was ever executed for three times the amount which this would cost.

For the first 8½ miles from the Chicago river, the material to be excavated is a compact clay, all of which can be easily excavated by machinery. This has been estimated at 25 cents per cubic yard.

For the next 10¼ miles, the excavation will be of a much more difficult character, though still mostly earth, but a considerable proportion of it cemented clay, intermixed with small stone or gravel. It is believed, however, that the greater part of it may be executed by machinery, though with less facility than the first 8½ miles. On a few of the sections embedded rock was found in the old excavations. This part of the work is estimated at 50 cents per cubic yard, including all the various kinds of material.

From the Sag, where the heavy rock-excavation commences, to Lockport, a distance of 10¼ miles, the excavation, except a slight covering of earth averaging about two feet in depth, consists entirely of stratified limestone. For some eight miles of this distance the depth of rock varies from 12 to 16 feet.

All of the excavation, on this part of the line, was completed on the original plan of the "deep cut," except about 260,000 cubic yards,—so that it only requires this amount to be excavated to make a perfect drainage to the bottom of the enlarged canal, and give a face the full depth of the excavation for the entire distance. Considering the character of the rock and the favorable circumstances for executing the work, (permitting the use of machinery propelled by steam for drilling and removing it,) we have deemed 90 cents per cubic yard a liberal price, and have estimated it at that price.

We mention, particularly, the character of the material to be excavated on this part of the work, and the prices at which we have estimated it, because its cost will be more than half of the entire improvement. But this cut through the Summit, though expensive, accomplishes a very important object. It diverts the waters of Lake Michigan into the valley of the Des-Plaines at Lockport, through a canal 160 feet wide at surface, and not less than 7½ feet deep in an ordinary stage of the Lake. A declivity of one inch per mile at the very lowest stage of Lake Michigan, has been given on this 29 miles of canal, in the estimates.

From Lockport, where the lake level runs out to the lower dam on the present canal at Joliet, the distance is 4½ miles, and the lockage 50 feet.

The whole amount of lockage from the point where we leave the present canal at Joliet to La Salle, is 88 feet, and from thence to the mouth of the Illinois river, 32 feet,—making the aggregate lockage from Lake Michigan to the Mississippi, 170 feet. The whole distance is 320 miles.

To overcome this difference of level we have estimated in all above La Salle, 14 lift locks,—the lifts varying from 8 to 12 feet each. But one guard-lock is required.

All but three of the nine locks below Joliet will be built upon short sections of canal, and entirely secure from river-floods.

Only five dams have been found necessary between Joliet and La Salle; two of them on the Des-Plaines river, and three on the Illinois. The two on the Des-Plaines have been estimated entirely of stone, and the three on the Illinois of crib-work, resting on timber foundations with stone abutments,—being of the same character in all respects as those which were estimated for the Illinois river improvement below La Salle.

Upon a re-examination of our estimate for the improvement of the river below La Salle, which we submitted to Gen. Webster, we conclude to make no change in the figures, as we deem them as near right as we could make them by revision now.

We have estimated the whole work to be constructed in the most substantial and permanent manner, but have added nothing for show. All the stone masonry, therefore, in locks, dams, bridges, etc., has been estimated of "rock-work," with a face sufficiently smooth for practical purposes, but little of it, except coping, finely cut. The stone in the quarries along the line is peculiarly fitted for this character of work, being regularly stratified, the strata being of various and suitable thicknesses, and the beds parallel and generally so smooth as to require but little dressing. All is to be laid in the best cement of water-lime and sand.

The improvement which we have estimated is not a "ship-canal," as it has generally been termed, it not being adapted to the navigation of *ships*, but *the largest class of Mississippi steamboats which can ascend the river in ordinary low water to St. Louis.* The canal, therefore, where the improvement consists of canal, is estimated throughout at not less than 160 feet wide at top water-line; and, where the canal is not made in rock-excavation, vertical walls upon each side, 10 feet in height, have been estimated.

The locks are estimated 350 feet long by 75 feet wide in the chamber, and would be sufficiently large to pass 12 of our ordinary canal-boats, carrying 2,000 tons of freight; or a gun-boat, 200 feet long by 40 feet wide, and drawing ordinarily 10 feet of water, could, by the aid of properly constructed lighters or barges, be buoyed up, so as to draw less than six feet, and thus be passed through from one end of the improvement to the other without the slightest difficulty.

The total cost, as will be seen by the annexed estimate, we have set down at the sum of $13,446,625, which amount varies but little from the estimate made by us last fall, before we were in possession of the data furnished by Mr. Mathewson's recent survey.

All of which is respectfully submitted.

WM. GOODING.
J. B. PRESTON.

Estimated Cost *of enlarging the Illinois and Michigan canal, from Chicago to the Des-Plaines river at Joliet, and improvement of the Des-Plaines and Illinois rivers to the Mississippi* —320 *miles in all.*

Chicago to Lockport—Summit Line, 29 Miles.

4,502,110	Cub. yds. earth excavation (to Summit),	@	.25	$1,125,527.50
5,931,612	" " and C. clay, etc. (to Sag.),	"	.50	2,965,806.00
2,521,189	" rock excavation (to Lockport),	"	.90	2,269,070.10
474,439	" earth " and spoil bank,	"	.15	71,165.80
512,152	" rock in "	"	.15	78,172.80
316,000	" vertical protection wall,	"	$2.50	790,000.00
				$7,299,742.20

Lockport to Dam No 2, in Joliet, 4½ miles.

187,360	Cub. yds. earth excavation,	@	.20	$37,472.00
198,619	" rock excavation,	"	$1.00	198,619.00
150,213	" embk't rock and wall to be remov'd,	"	.20	30,042.60
66,195	" vertical protection wall,	"	2.00	132,390.00
3,000	" lining for banks,	"	1.00	3,000.00
	5 locks, 350 ft. long, 75 ft. wide, 10 ft. lift each,			474,800.00
				$876,323.60

Dam No. 2, Joliet to Lake Joliet, in Des-Plaines River, 3½ Miles.

25,925	Cub. yds. rock excavation,	@	$1.00	$25,925.00
10,500	" mortared wall in embankment,	"	3.00	31,500.00
189,201	" embankment east side of river,	"	.30	56,760.30
191,250	" " through fields,	"	.20	38,250.00
19,800	" protection wall,	"	$2.50	49,500.00
	1 stone dam, 350 ft. long, 15 ft. high,			17,050.00
	3 locks, 350 ft. long, 75 ft. wide, 8 ft. lift each,			281,100.00
				$500,085.30

Improvement of Des-Plaines and Illinois Rivers from Lake Joliet to La Salle, 59 Miles.

59,259	Cub. yds. rock excavation at Treat's Island,	@	$1.00	$59,259.00
620,000	" earth "	"	.25	155,000.00
395,917	" " and C. clay,	"	.30	178,775.10
91,584	" rock excavation,	"	.75	68,688.00
250,000	" embankment,	"	.25	50,000.00
800,000	" earth excavation and embankment,	"	.30	240,000.00
108,186⅔	" protection wall,	"	3.00	324,560.00
33,000	" slope wall,	"	2.00	66,000.00
	1 dam in Des-Plaines river,			24,190.00
	3 dams in Illinois river,			205,660.00
	2 locks, 8 ft. lift each, 350 ft. long, 75 ft. wide,			213,700.00
	1 guard lock, " "			93,700.00
	4 locks, 12 ft. lift each, " "			519,400.00
				$2,198,932.10

La Salle to the Mississippi River, 220 Miles.

7 tree or timber and crib dams, with stone abutments,	$516,040.00
7 stone locks, 350 ft. long, 75 ft. wide, 32 ft. lockage,	1,128,295.00
	$1,644,335.00

RECAPITULATION.

Chicago to Bridgeport, South Branch,		4 miles.	
Bridgeport to Lockport, Summit line,		29 "	$7,299,742.20
Lockport to dam No. 2, Joliet,	50 ft. lockage,	4½ "	876,323.60
Dam No. 2 to Lake Joliet,	24 " "	3½ "	500,085.30
Lake Joliet to La Salle,	64 " "	59 "	2,198,932.10
La Salle to Mississippi river,	32 " "	220 "	1,644,335.00
Add for bridges, 2 culverts, and land damages,			325,000.00
" engineering and contingencies,			602,207.10
Totals,	170 ft. lockage,	320 miles,	$13,446,625.30

[C.]

NIAGARA SHIP-CANAL.

This is a project which has been discussed for more than a quarter of a century; and has recently received especial prominence as an additional Eastern outlet for the accumulating products of the Mississippi valley.

In 1835, Capt. W. G. Williams, under the orders of the Chief of the Topographical Bureau of the United States, made a survey of the proposed route, of which the following is an abstract:

The project contemplated a ship-canal of such dimensions as to render the work a means of transportation for the larger class of steam-boats and sailing vessels navigating the lakes; with locks 200 feet in length, 50 feet in breadth, with a width of canal of 110 feet at the surface, and a water-depth of 10 feet:—the locks to have a lift varying with circumstances, and generally not to exceed 10 feet, and the water to supply them to be drawn from the Niagara river.

PROPOSED ROUTES.

No. 1. Commences at Porter's Store-house near old Fort Schlosser, thence crosses Gill creek at a distance of half a mile above its mouth, and is thence protracted in a nearly straight line to the head of Bloody run. The ground over which it passes, after the first mile, is generally swampy, although somewhat elevated, and for the first four miles, as determined by careful borings, no rock worthy of mention occurs, except a small portion at Gill creek; the soil, however, is by no means easy of excavation, being of a tenacious character; and the ground is swampy, and covered with a heavy growth of timber. "From this point," quoting from the Report, "the valley of Bloody run is pursued to within a short distance of the point, where the run falls over the precipice into the Niagara river, at a short distance from the chasm known as the Devil's hole, three and a half miles below the Great Falls.

"The levels now pass over unequal ground, but slightly elevated, however, until they reach the brow of the Lewiston ridge. This portion of the line was run very near the precipitous brink of the Niagara river, and only involves a prism of rock cutting of inconsiderable depth.

"Until we arrive at Fort Grey, no obstacle of importance intervenes; indeed, none but the most commonplace circumstances of canal construction present themselves. It is from this point to the debouch of the project into the Niagara river, that difficulties of a serious character may be apprehended.

"From the brow of the ridge the lines of level were carried

obliquely to the line of greatest acclivity of the ascent, falling in such proportion to the measured horizontal distance, as to render them conformable to the projected dimensions of the locks and basins, with the required lift for each lock. These data furnish the means of projecting a flight of double consecutive locks to the foot of the ridge, or a line of single locks, with intermediate basins, involving in either case a descent of 319½ feet from the bottom of the canal at Fort Grey, to the corresponding surface at its intersection, ten feet below the surface in Niagara river.

"As the slope of the mountain may in a general view be regarded as uniform, and under an angle too great to admit of the location of the locks on a line approximating to that of greatest acclivity, it would be necessary, by means of excavation and embankment, to prepare a berm for their reception.

"Our supposition involves a heavy mass of side cutting, so as to establish the exterior walls of the locks upon a well consolidated foundation: by this means the whole section of the locks and basins would possess a homogeneous basis, and have their stability insured.

"This excavation comprehends the space to be occupied by the sustaining and interior walls, and in case the double locks should have their similar surfaces in the same horizontal plane, the breadth of their dividing walls would be comprised in the section. * * *

"As the line of levels descends to the foot of the ridge, it gradually winds round until its horizontal projection becomes nearly parallel to its location at the beginning of the descent.

"In order to obtain the direction which leads it to the most favorable point of debouch on the Niagara river, for the present modification of our project, I have planned a basin allowing sufficient room for the largest vessels admissible to the locks, to turn and assume its change of course. At this point, the flight of locks would terminate in an extensive artificial harbor, comprising an area of about 114 acres, and elevated 120 feet above the level of the Niagara river. It will be formed between the ridge on which the principal street of Lewiston is situated and the main ridge, possessing a mean depth of fourteen feet. The embankment necessary to back the water would be very inconsiderable.

"It is an element forming a very important feature in our project, and would have the advantage of serving as a part of the canal, obviate a mass of expensive construction, and at the same time afford very essential accommodation to trade; indeed, a basin of this kind would be almost necessary by reference to the very contracted space which can be made available for the purpose of commercial transactions in the vicinity of the debouch, in connection with the precipitous banks of the river, and the violence of the current; moreover, the prism of water drawn from this reservoir to supply the descent of the locks to the termination of the project, would be scarcely perceptible. This would render the descent from the harbor to the outlet, independent for its immediate exigences, of the supply of water, to be drawn through the upper flight of locks from the summit level of the project."

Elaborate estimates are given as to the cost of construction on this plan, with the following results:

Total amount of Lockage	$2,852,208.58
Cost of Canal	758,387.63
Total	$3,610,596.21

Total length, 7 miles, 4,040 feet.

No. 2. "The projected plan, line No. 1, approaches in a great portion of its development to the frontier of Upper Canada, and it is in this part clearly within the range of howitzer and mortar batteries, planted on the opposite shore of the Niagara river, and likewise entirely under the influence of their power of annoyance at its outlet. And this circumstance has been alleged against the expediency of any location terminating at Lewiston. In order to remove any objection which indeed is valid, by reference to the military advantages that are ascribed to the undertaking, a route was examined by which the inconveniences referred to might be avoided.

"A line of levels was accordingly carried up the valley of Gill creek, to a depression which occurs in the Lewiston ridge at the head of Fish creek, and thence descending the ridge, following the valley of Four-mile creek, to its termination on Lake Ontario, and conforming in general direction to a right line between its two extremities.

"This route fulfills the condition required. It is throughout its development entirely without the range of annoyance from the opposite shore of Niagara river, and terminates on the lake in deep water.

"It must be remarked, in regard to this location, that it exacts very deep cutting in rock for a distance of 3½ miles, but that in other respects the difficulties to be surmounted are less than upon any other route, and particularly in that part which relates to the descent of the ridge. The idea also of the expense in excavation, should be neutralized by the reflection, that the rock taken out would be of essential value for the construction of the harbor, locks, walls, etc.

"It is probable that the whole amount of stone taken from the deep cut would be required for these and a variety of contingent purposes, but more particularly for the construction of a capacious and efficient harbor at the debouch of the project.

"For such object, the stone would be required from some point or other, and from none could it be procured at so cheap a rate as from the excavation in question.

"The harbor would be the last object of completion, and the stone conveniently deposited on the side of the deep cut, would be brought down by the boats through the canal. By means of this abundant supply of materials, the wharves might be carried out to a depth of fourteen or fifteen feet at little expense, so as to avoid inconvenience from alluvial deposits."

The cost of construction of this line, including an artificial harbor at the mouth of Four-mile creek, with a circuit of one mile, with piers from 7 to 10 feet above the surface of the water, and 30 feet cross section, would be $4,616,423.47.

Total length, 14 miles, 5,000 feet.

There are several modifications of the above routes, in the surveys, but it may be said that route No. 1, if constructed for purely commercial purposes, is the shortest and least expensive; but when the question becomes one of military defense, it is apparent that route No. 2, is the one which ought to be adopted, as affording security from assault, and incalculable aids to the national defense.

The dimensions of the work, however, are too contracted, in view of the number and the class of vessels navigating the Lakes. The locks ought to be 350 feet long, 70 feet wide and 12 feet deep; while the bottom of the canal ought to be 160 feet wide; or 100 feet with recesses of 160 feet in width, at each mile, to admit the passage of vessels.

The lower portion of this route may be regarded as one of the most defensible positions on the frontier, inaccessible by means of the rushing waters and the precipitous banks of the Niagara river; but above Schlosser, the river-channel is in part on the Canadian side, and, therefore, Capt. Williams proposed to pass the rapids at Black Rock, by a short cut and a few feet of lockage on the American side. Another suggestion of his was, to let the canal commence at Buffalo, and carry the level of the lake along the margin of the Niagara river, as far retired as possible to the mouth of Gill creek, then up the valley of this stream to the head of Fish creek, as per route 2, and thence descend to Lake Ontario. This modification, while adding to the expense, would give additional security, and at the same time save a very considerable prism of rock-excavation, proportionate to the elevation of Lake Erie above the Niagara river at the point of beginning, near Porter's Storehouse.

The construction of this work, while it would be the grandest monument of engineering skill on the continent, at the same time would vastly contribute to the military defense and commercial facilities of the country. (*Rep. of Capt. W. G. Williams, U. S. Top. Eng., Doc.* 214, *Ho. of Reps.*, 24*th Cong.*, 1*st Session.*)

[D.]

RAPIDS OF THE UPPER MISSISSIPPI.

Serious impediments exist in the navigation of this river, in consequence of two rapids, known as the Upper, or Rock Island, and the Lower, or Des-Moines, Rapids. The foot of the Lower Rapids commences about one-half mile above Keokuk, Iowa, from which they extend ten and one-half miles, and are occasioned by four reefs of limestone, stratified and nearly horizontal in position, which are known as the Lower, English, La Malle's, and Upper chains. The first three are nearly contiguous, while the last is 2½ miles distant, the interval being comparatively unobstructed.

The fall from the head to the foot of the obstructions, at low water, is 21 feet; the slope, however, is not uniform,—the maximum being 6 feet per mile, while the minimum is only about $\frac{2}{10}$ per mile.

The minimum range of the water-surface at the head of the rapids, is 11¾ feet, and at the foot 21 feet. Hence, the high-water fall is 13¼ feet: and the average fall 1.12 foot per mile. The velocity at low-water nowhere reaches five miles per hour; but at high water it reaches probably seven. Steamboats of the least tonnage, drawing two feet of water, cannot pass at low-water without imminent danger of getting fast on the rocks, and navigation is entirely suspended, at this stage of water. According to the report of Lieut. Warren, of the Topographical Engineers, (now Brigadier General, and Chief of Staff to the Commander of the Army of the Potomac), there is three feet of water on the bars above and below the rapids; and as the water rises, owing to the greater width and slope of the rapids, the water increases on the bars more than on the reefs, in about the following ratio:

With	3	feet on	the bars,	there would be	4	feet on the rapids.
"	4	"	"	"	4¼	" "
"	5	"	"	"	4½	" "
"	6	"	"	"	5	" "
"	7	"	"	"	5⅜	" "
"	8	"	"	"	6½	" "

In 1837, Lieut. R. E. Lee, now Major General of the Confederate army, was detailed by the Chief of the Topographical Bureau to take charge of the operations for the improvement of the navigation of these Rapids. His system consisted in deepening the channel of the water by blasting out the rocks; and the result was, according to the testimony of pilots, that steamboats drawing from 9 to 12 inches more water, were enabled to pass. His operations were confined to the Lower and English chains. The Omega patch, or Lower chain, is regarded as the most difficult of all to navigate, owing to swift cross-currents; but the obstructions would have been removed, had the work been continued.

The following is the estimated cost of improving the natural channel of these Rapids:

FOR CHANNEL 100 AND 200 FEET WIDE RESPECTIVELY, AND 4 FEET DEEP.

	100 feet. Cubic yards.	200 feet. Cubic yards.	Cost 100 feet. $10 per yd.	Cost 200 feet. $10 per yd.
Upper Chain	3,710	14,070	$37,100	$140,700
Nashville Crossing	1,288	2,777	12,880	27,770
La Malle's Chain	7,927	32,897	79,270	328,970
English Chain	4,243	21,165	42,430	211,650
Lower Chain	4,813	18,444	48,130	184,440
Total	21,981	89,353	$219,810	$893,530

UPPER, OR ROCK ISLAND RAPIDS.

These rapids commence about one-half mile above the lower end of Rock Island, and extend for 13 miles up the river. The principal reefs are known as the Lower chain; Rock Island chain, two miles above; Duck-creek chain, four and one-half miles above; Campbell's chain, seven and one-half miles above; St. Louis chain, ten miles above; Sycamore chain, twelve miles above; and Upper chain, at the head of the rapids.

The rocks consist of friable limestone, easily quarried, and a soft yellow sandstone, with occasional granite boulders. Small steamboats, drawing 2½ feet of water, pass at the lowest stages.

Duck and Campbell's chains are particularly dangerous, as a boat, to avoid the rocks, is compelled to make abrupt turns.

Rock Island chain is a continuous flat reef across the river-bed, with a low-water depth of 2½ feet. By connecting the islands with the Illinois shore by a dam, the water has been raised about ten inches; and, at the same time, the width of the river has been reduced to 400 yards. The average width of the rapids is one-half mile.

The fall of the whole series of rapids, at lowest water, is 22 feet, with an average slope of $1\frac{7}{10}$ foot per mile. The greatest slopes of surface are at Upper, Sycamore, and Rock Island chains, the current being between four and five miles per hour. The range from low to high water at the head of the rapids is 13 feet, and at the foot 23 feet, making the high-water fall 12 feet, or $\frac{12}{13}$ foot per mile.

The following is the estimated cost of the removal of rock:

FOR A CHANNEL 100 AND 200 FEET WIDE RESPECTIVELY, AND 4 FEET DEEP.

	100 feet. Cubic yards.	200 feet. Cubic yards.	Cost 100 feet. $10 per yd.	Cost 200 feet. $10 per yd.
Upper Chain	2,960	7,095	$29,600	$70,950
Sycamore Chain	560	7,770	5,600	77,700
Crab Island		1,300		13,000
St. Louis Chain	275	1,540	2,750	15,400
Campbell's Chain	2,110	9,380	21,100	93,800
Winnebago Island	180	180	1,800	1,800
Duck Creek Chain	1,988	3,815	19,880	38,150
Rock Island Chain	3,890	8,200	38,900	82,000
Lower Chain	400	5,735	4,000	57,350
Total	12,362	45,015	$123,620	$450,150

"The practicability," remarks Lieut. Warren, "of improving the channel by removing the rocks in the bed of the stream has been proved beyond question, by the results of former operations; and a careful examination of the obstructions themselves is sufficient to convince any one without further demonstration. All of the rapids pilots I have heard on the subject, are unanimously of this opinion; so no apprehension need be felt, that by enlarging the narrow passes we shall lower the water in the pools above, and cause new obstructions. The effect of a sluice-way, two hundred feet wide, cannot sensibly affect the pools, with a whole river for their supply; and even if so, it can be remedied by depositing the excavated rock in the numerous small chutes. Sharp edges will, in most cases, be avoided at the sides of the channel, as the improvement will be effected by removing narrow ledges that, at present, divide the channel, and the sides will remain as before."

The blasting, in the operations heretofore performed, was effected as economically and as rapidly as on shore, the only drawback having been the limited number of days of low-water in which the work could be prosecuted, amounting to 28 days in the fall of 1828, and three months in 1839. The detriment which these two rapids occasion to commerce is estimated by Mr. O'Sullivan, Civil Engineer, at $600,000 annually. (*Report of G. K. Warren, Lieut. Topographical Engineers U. S. A. to Supt. of Western River Improvements, Ex. Doc.* 104, 33*rd Congress,* 1*st Sess., House Reps.*)

[E.]

IMPROVEMENT OF THE FOX AND WISCONSIN RIVERS.

Commencing at Green Bay, the line of this route is up the valley of the Fox river, forty-six miles to Lake Winnebago; thence through this lake for sixteen miles, to the mouth of the Upper Fox; thence by that stream, 110 miles to Portage City, formerly Ft. Winnebago; thence by artificial canal, two miles, into the Lower Wisconsin; thence down that stream to its junction with the Mississippi, 115 miles—making, in all, a distance of 283 miles. This route has been partially improved by a private corporation, under the authority of the State of Wisconsin, but the navigation is precarious, steamboats being able to ascend the Wisconsin only during the spring and fall freshets.

"The descent of the Lower Wisconsin," says Mr. Jenne, the former Superintendent of this improvement, "is about one foot to the mile; it is from 500 to 1000 feet wide, with a current of two miles per hour. There is a channel, in all cases, of five to six feet deep, in low water; but this being crooked when the water spreads out, requires to be reduced in width, by means of wing-dams, when the river will make it own channel as it recedes from high to low water."

The cost of improving the Lower Wisconsin, 115 miles in length, to a permanent depth of at least six feet, is estimated by Mr. Jenne at $250,000. The improvement of the Fox river, and a canal at Portage City, have been made by locks and dams, and by dredging, by which, a permanent depth of four feet has been secured, from Portage City to Green Bay. The cost of enlarging the locks, so as to admit a boat 200 feet long, and 34 feet beam, and deepening the channel to six feet, at the lowest stages, is estimated, by the same engineer, at $1,000,000, and the time required to complete the work, two years.

Lieutenant Colonel Cram, of the Topographical Engineers, has proposed Lake Winnebago as a naval station for the Lakes. This lake is twenty-eight miles long, and ten broad; its circumference is 74 miles, and it covers 212 square miles; and the depth of water off shore of only 600 feet, is from eight to eleven feet. Its eastern shore is bordered with a series of bluffs of limestone, which affords good material for construction, and at the same time, readily calcines into lime. Ship-timber, consisting of pine and oak, abounds in the immediate vicinity. The soil is good, and the climate healthy and invigorating. But the surface of Lake Winnebago is elevated 164 feet above that of Lake Michigan, and to enable a war-vessel 200 feet long, by 34 feet beam, and requiring a depth of water of 12 feet, to pass from one lake to the other, would involve the construction and enlargement of some twenty locks, and the dredging of 24,000 cubic yards of silt at the mouth of Fox river—the whole expense of which, as estimated by Col. Cram, would be $1,662,384.

There are sites for a naval station on the borders of Lake Michigan, which, while they possess all of the advantages enumerated by Col. Cram, are not subject to this serious disadvantage, of requiring a lockage of 160 feet, in a distance of 46 miles to reach the Lakes. For example, between Little Traverse bay and the Michigan shore opposite Bois Blanc Island, there is a chain of lakes, deep and high-banked, some of which are from 10 to 15 miles in length by 5 in breadth, and elevated but a few feet above the surface of the lake. Grand Traverse bay communicates with a similar series of inland sheets of water. A short passage communicates with Elk lake, which is connected by an inconsiderable channel with Torch lake, which is 20 miles long and 8 miles broad, with a depth of over a thousand feet, while the harbor of the bay is secure and capacious. The height of this chain is but eight feet above the bay, so that while a single lock would admit of the passage of a vessel, at the same time would afford an unlimited water-power, at a point where it would be required. The head of Torch lake approaches within one-half mile of the bay, between which the excavation of a canal is entirely practicable. The series of islands known as the Manitou, Fox, Beaver, Garden, and Hog Islands, afford shelter in every storm, while the spacious bays, Grand and Little Traverse, have yielded the deepest soundings yet attained in Lake Michigan.

Mackinac, or rather the opposite points, Gros Cap and La Barb, as known to every one familiar with the Geography of the Lakes, is the key to Lake Michigan, and in a measure to Lake Superior. What Gibraltar is to the Mediterranean, what Constantinople is to the Bosphorus, what Quebec is to the St. Lawrence, Mackinac may be made to the North-West; and the power which holds this important passage can close it against any naval enemy, and afford ample protection to the fleets of commerce.

As preliminary, therefore, to all defenses of the Northern Lakes, Mackinac should be made impregnable. A navy yard, at either point mentioned, would be rendered secure by the protecting guns of the Straits, and at the same time, the floating armaments would be within supporting distance;—ready to aid in repelling attack, or to float into Huron and Erie.

The cost of such a work would be far less than that suggested by Col. Cram, and would be within cannon range of the waters on which the vessels were designed to float. (*Communication of Lt. Col. T. J. Cram; Senate; Miscellaneous Doc. No.* 14, 37*th Congress*, 3*rd Session.*)

[F.]

ENLARGEMENT OF THE NEW YORK CANALS.

By a resolution of the Assembly of that State, passed March, 1862, the State Engineer and Surveyor was directed to examine and report on the practicability of the enlargement of one tier of locks through the length of the Erie and Oswego canals, to 150 feet in length and 25 feet in width, for the consideration of the General Government, as connected with the subject of lake and frontier-defense.

In obedience to said resolution (Assembly Document No. 8), the State Engineer reported the cost of enlarging the canals to the above dimensions, so as to admit the passage of iron-clad gunboats of the dimensions therein mentioned, as follows:

Enlarging the locks on the Erie canal	$2,815,900.00
Enlarging the locks on the Oswego canal	625,500.00
	$3,441,400.00

The Committee of the Assembly thus speak of this proposed enlarged communication:

"The enlarged Erie canal is of itself a splendid, though artificial river; its depth of water is seven, and its breadth seventy feet. It is reliable, constant and well-protected, and on its peaceful bosom is borne a vast tonnage, immense wealth, and an almost imperial commerce. True, it is and was designed as a great commercial channel, a highway for the products, whether agricultural or manufactured, of a vast country, to the best markets of that country and of the world; but it seems eminently proper that the ways and means thus furnished for the purposes of peaceful trade should be converted, in case of necessity, into a means of defense and protection to that trade. It will be an era in the history of the nation and of public expenditures, when money expended for purely defensive purposes shall, in the largest and best sense, serve the ends and multiply the facilities of internal trade."

By an act of the Legislature of New York, passed April 22, 1862, it is stipulated that,

"Whenever the Government of the United States shall provide the means, either in cash, or six per cent. stock or bonds, redeemable within 20 years, for defraying the cost of enlarging a single tier of locks, or building an additional tier, in whole or in part, upon the Erie and Oswego canals, including any necessary alteration of said canals, or other structures, to a size sufficient to pass vessels adequate to the defense of the Northern and North-western Lakes, the Canal Board shall, without delay, put such work under contract, in the manner required by law, to be constructed and completed at the earliest practicable period, without serious interruption to navigation; with power in the discretion of the Canal Board to direct the construction of new and independent locks, when found more advantageous."

"On completing the said work on either of the said canals, the Government of the United States shall have the perpetual right of passing through the canals thus enlarged or built, free from toll or charge, for its vessels of war, boats, gun-boats, transports, troops, supplies or munitions of war, subject to the general regulations prescribed by the State, from time to time, for the navigation of its canals."

Inasmuch as the estimates of the State Engineer were assailed by the Military Committee of the House of Representatives of the Third Session of the 37th Congress, the General Assembly, at its last session, directed a re-survey of the said canals, by the State Engineer and a United States Engineer, with a view of adapting them to naval defense, and in reference to the cost, amount of water, etc. etc., which information will be laid before the Assembly at the next session.

The memorial of the State Agent, the Hon. S. B. Ruggles, addressed to the President of the United States, and by him communicated to Congress, and the memorial of the New York Chamber of Commerce, transmitted to the Senate of the United States, at the last session, contain a fund of statistical information as to the resources of the country, and the necessity of enlarged outlets, which is worthy the attention of every legislator.

In this connection, is presented the subjoined comprehensive letter of the Hon. N. S. Benton, Auditor of the Canal Department of New York, in reference to the cost, financial condition, the amount of freight, and the lockage capacity of the Erie canal.

Canal Department,
Albany, *May* 29, 1863.

Dear Sir: Heartily concurring in the objects expressed in the call for a National Convention to assemble at Chicago on the 2nd day of June next, and being unable to attend, I take the liberty of submitting some remarks relating to the subject that will be brought before the Convention for consideration and discussion. I may say now that the points embraced in this communication are suggested by the resolution of the Honorable Canal Board of this State, a copy of which is annexed to this letter. They are—

1. A statement of the cost of the State canals connecting the waters of the Hudson river with the Northern and Western lakes.

2. A view of the present and prospective financial condition of the State relating to her canals. This is given to show her present inability to contribute materially as a State towards the objects contemplated by the Convention.

3. A statement of the shipment of freights at the western termination of the Erie and the northern termination of the Oswego canals in 1860, 1861 and 1862; with the tide-water deliveries by the Erie canal of Western and Canadian products, during the same period; also showing the large annual increase both of shipments and deliveries.

4. A statement of the present lockage capacity of the Erie canal. The enlargement of this canal has been declared completed. Some improvements in its water-channel may be needed to give boats six feet draft of water, and an abundant appropriation has been made by the Legislature, and now awaits expenditure by the Canal Commissioners.

1. *Of the cost of construction of the Erie, Champlain, and Oswego canals, and of the enlargement of the same.* The cost of the other State canals is purposely omitted, as they have no direct connection with the trade of the Western States and Canada.

Cost of construction and completion of the Erie and Champlain canals, and of the enlargement of the Locks on the Champlain canal	$12,195,699.63
Cost of the enlargement of the Erie canal, commenced in 1835...	31,284,696.39
Cost of original construction, and of the enlargement of the Oswego canal	3,077,429.57
	$46,557,825.59

In this statement I have not included the interest paid on loans for money borrowed for the construction, enlargement, and completion of these works. Nor have I brought into account the annual expenditures and outlays for superintendence, collection of tolls, and the ordinary repairs and maintenance of these canals.

2. *Of the present canal debt of the State;* the reimbursement of which is, under the constitution of 1846, charged directly upon the canal tolls and revenues; and it is made the duty of the State officers to levy a sufficient toll upon all boats and property passing on the canals, to meet the annual charges imposed by the constitution, including the charges for ordinary repairs.

1862. Sept. 30. Canal debt contracted prior to 1846, for the construction of the canals..	$8,839,024.76
The canal debt contracted since 1846 and prior to 1860, for the enlargement and completion of the canals..................	12,000,000.00
Total funded debt, charged on canal revenues..........	$20,839,024.76

The annual contribution required by the constitution to be made from the canal tolls and revenues, to pay the interest and reimburse the principal of the above debt as it falls due, amounts to	$2,816,242.63
In addition to the above, there is a further annual constitutional charge upon these tolls, to reimburse the treasury, for the sum of $8,032,995,702, advanced by the Treasury on account of the canals before 1846, of..	550,000.00
Total annual charge...............	$3,366,242.63

If the canal-revenue sinking funds are at any time deficient, the constitution requires an annual levy of direct taxes to make good such deficiency.

To the above must be added a further debt,* contracted for canal purposes upon the credit of the State, prior to 1854, to be reimbursed by direct taxation, of..........................	$442,585.49
And also a further debt contracted for like purposes in 1859, of..	2,500,000.00
Floating unfunded debt for claims due to contractors, and for land damages..	1,500,000.00
	$4,442,585.49
The annual charge upon the treasury, to be supplied by direct taxation to pay the interest and reimburse the principal of the debt, so much of it as has been funded, amounts to................	$313,518.28

* $200,000 of this debt, as it stood at the close of the last fiscal year, due next October, is omitted, as we have funds on hand to pay it.

RECAPITULATION.

	DEBT.	CONTRIBUTION.
Stock-debt of the State, chargeable upon canal revenues........	$20,839,024.76	
Required annual contribution...........		$3,366,242.63
Stock and floating debt chargeable upon the treasury...............	4,442,585.49	
Required annual contribution, paid by direct taxation...........		313,518.28
Totals..............................	$25,281,610.25	$3,679,760.91

This annual contribution is in addition to the annual appropriation of $800,000, for superintendence, collection and ordinary repairs.

Since 1853 we have levied and collected direct taxes, and applied the same to the enlargement and completion of our canals, to the amount of..	$4,800,671.99
Since 1854 we have raised taxes to supply the deficiencies in our canal sinking funds to pay interest, to the amount of.........	3,318,448.21
Raised by taxes to make good canal-revenue deficiencies for the General Fund..	2,737,500.00
Total taxation for canal purposes in nine years.........	$10,856,620.20

We have no available or reliable means to pay off our present canal-floating debt of $1.500,000 due to our citizens, but a resort to the pockets of our tax-paying citizens.

The constitution, in providing for the payment of our canal-debt, makes no distinction between the debt contracted for the enlargement and one incurred for the construction of the lateral canals.

3. *Of the shipments of Western and Canadian products.*

Shipments of freight at Buffalo in 1860, were............	1,113,754	tons.
" " Tonawanda.....................	116,502	"
" " Oswego.........................	751,900	"
Total..	1,982,156	tons.

In 1861 there were cleared at Buffalo...................	1,579,745	tons.
" " " Tonawanda...............	77,036	"
" " " Oswego..................	609,222	"
Total	2,266,003	tons.

In 1862, cleared at Buffalo	1,980,945 tons.
" " Tonawanda	99,711 "
" " Oswego	638,419 "
Total	2,719,075 tons.

Increase in 1862 over 1860	736,919 tons.
Increase in 1862 over 1861	453,072 "

The tide-water deliveries by the Erie canal from the Western States and Canada in 1860, were	1,896,975 tons.
1861	2,158,425 "
1862	2,594,837 "

Increase in 1862 over 1860	697,862 tons.
Increase in 1862 over 1861	436,412 "

I will now present the tide-water deliveries by the Erie canal from the Western States and Canada for several years:

In 1842	221,477 tons.
" 1847	812,840 "
" 1852	1,151,978 "
" 1857	919,998 "
" 1862	2,594,837 "

Showing an increase of Western tonnage delivered at tide water since 1842, equal to $1171\frac{60}{100}$ per cent., and since 1852, equal to $225\frac{25}{100}$ per cent.

The average cargo of boats in 1842, was	42 tons.
" " " 1847	65 "
" " " 1852	80 "
" " " 1857	100 "
" " " 1862	167 "

Whole number of lockages at Alexander's Lock during 1842, was	22,869.
" " " " " 1847,	43,957.
" " " " " 1852,	41,572.
" " " " " 1857,	22,182.
" " " " " 1862,	34,977.

The rate of increase on the average of the cargoes of boats from period to period, is 55 per cent., 23 per cent., 25 per cent., and 67 per cent.

4. *Of the lockage capacity of the Erie canal.* I now give my views of the capacity of the enlarged Erie canal, operated with single and double locks of the enlarged size.

I assume the canal completed, with a water way of 70 feet by 7, giving to canal boats of the largest class six feet draft of water, and a capacity to carry 7,000 bushels of wheat, 210 tons. We can pass a boat through a single lock in ten minutes, or make 144 lockages in twenty-four hours—72 lockages each way, down and return, per day. During the season of navigation, 225 days, we can pass through a single lock, 14,750 loaded boats, of 200 tons cargo each, and this gives an aggregate for the season of 3,240,000 tons.

This estimate is made on the assumption that the lock is in good working order every day, that there are no breaks or detentions, and that the locks are well and properly attended by good lock tenders.

Double locks will not perform double the service of a single one. But double locks have the capacity of making 40,000 lockages in a season of navigation, or 20,000 each way. Some of our double locks have made 43,957 lockages in a season. On this basis the capacity of the double locks is 4,000,000 tons each way. An actual test would make the capacity of the double locks over the single locks larger than I have made it.

The whole number of tons delivered at tide water from the Erie canal in 1862, was 2,917,094, including the tonnage from the Westen States, etc. The whole number of tons delivered in 1859, was 1,451,333; so that a prospective increase for the next three years equal to that from 1859 to 1862, will bring up our canal tonnage to the full capacity of our double locks.

The entire limit of the working of the double locks at Alexander's would have been reached in 1862, with the tide water deliveries of 2,917,094 tons in boats carrying only 100 tons. The required lockages each way would have been 29,170 for the season, and 260 lockages per day.

A recurrence to the shipment and delivery of Western products before noticed will show how rapidly we are approaching our maximum.

Before we reach that point, however, our canal-debt will be so far reduced as to release our canal tolls from the yearly contribution of $1,700,000, when we shall be enabled to make a material reduction in our rates of toll.

MISCELLANEOUS REMARKS.

We have never imposed rates of toll for the express object of raising a surplus in excess of the constitutional charges upon the canal-revenues. We have obtained those surplus remainders at different periods, as the mere results of a prosperous season's traffic.

The rates of toll were so much reduced from those of 1857, by the action of the Canal Board and Legislature in 1858 and 1859, that the whole gross receipts from 3,781,684 tons of property carried in 1859, was only $1,723,915. Here we encountered a constitutional deficiency in revenue of more than $2,500,000. This state of things—lasting four years, with a prospect of continuance—would have carried the Erie canal even under the auctioneer's hammer. Better councils than those of 1858–9 prevailed in 1860. We saw, or thought we saw, the State would lose the canals, or we must take measures to restore the toll rates of 1857. By the toll-sheet adopted in the winter of 1860, we made a partial approach to the rates of 1857 from those of 1859. By the toll-sheet adopted before the 4th March, 1861, the rates of 1857 were attained on all Western products, except half a mill on wheat and flour. The rates on corn and corn meal were, in 1861, raised half a mill above those of 1857. In the winter of 1862 the rates on wheat and flour were brought to the standard of 1857. The rates of 1862 and 1863 are alike. The increase on wheat and flour in 1862 is no compensation to the State for the difference in the commercial value of the currency we receive for tolls, compared with the currency of 1857.

Although not exactly in place, perhaps, it is just and proper, and within the scope of the authority under which I make this communication, that I should notice certain published allegations that materially affect the honor and dignity of our State.

It has been asserted that we recently raised our rates of toll, and thus increased, by State authority, exactions upon Western products, in consequence of the free navigation of the Mississippi river being closed to the loyal people of the Western States. Was that navigation closed before the 4th of March, 1861? The free navigation of that river was obstructed by an act of the rebel Congress in February, 1861, and the first hostile gun was fired in Charleston harbor in April following. Neither the fact of obstruction, nor the anticipation of it, had any influence whatever in fixing the rates of tolls upon our canals.

Whatever of our wealth and means, be they much or little, or whatever of men or material may be required to open the Father of Waters to the free occupancy and enjoyment of our loyal brethren of the Western States, New York does not hesitate to give freely; when she does, then she may be justly charged, but not before, with imposing exactions upon the necessities of her loyal neighbors. This charge against the State is unkind, undeserved, and not justified by the facts of the case.

There may have been just causes of complaint in respect to the high rate of freight-charges the two past years; and if so, the following exhibit may explain to what extent this State and the carriers have participated in these high rates.

Charges by the Buffalo Route.

	c. m. fr.
In 1861 the highest monthly freight average on wheat from Chicago to New York, per bushel, was	$0 38 9.4
Lowest monthly average	0.17 2.5
Average for the season	0.27.2 8
State canal tolls, $0.05.1.7 per bushel.	

	c. m. fr.
1862. Highest monthly average	$0 34.9.4
Lowest monthly average	0.20.3.1
Average for the season	0.26.3.3
State canal tolls, $0.06.2.1 per bushel.	

For these toll-charges, New York furnished 345 miles of artificial navigation.

The freight charges by the Oswego route were very nearly the same both years.

In closing this communication, I have only to repeat here an opinion I have long entertained, that if we desire to increase the facilities of transit from the Lakes to the Atlantic by the way of the New York canals, we can only do so by enlarging the locks, so as to permit the passage of vessels through the present water-channels adapted to the use of steam entirely, and of sufficient capacity to carry 18,000 or 20,000 bushels of wheat. Any change from six feet draft of water, and an elevation of the bridges over twelve feet above the present surface line of the waters of the canal, would be impracticable, where the canal passes through our cities and villages. No satisfactory estimate ever can be made of the cost of such change.

The necessities of a reduction in the cost of transit on Western products to the Atlantic are too apparent to require examination or discussion; and it appears we have now reached a period when we should initiate measures with a view of providing facilities for the transport of the increasing products of the Western and North-western States.

I hope and trust these great objects will be accomplished by the wisdom and forecast of the Convention, about to assemble at Chicago.

I am very respectfully, etc.,

N. S. BENTON.

J. W. Foster, Esq.,
Secretary at Large, Chicago.

At a meeting of the Canal Board of the State of New York, held at the Canal Department, the 25th day of May, 1863.

Present: The Lieutenant-Governor, Comptroller, Secretary of State, State Engineer and Surveyor, and Canal Commissioners Alberger, Wright, and Skinner.

The Comptroller offered the following preamble and resolution, which were adopted:

Whereas, The Canal Board has been informed that the Governor has designated the Hon. Nathaniel S. Benton, Auditor of the Canal Department, as one of the delegates from this State to the Convention to be held at Chicago, on the 2nd day of June next, for the purpose of considering the subject of obtaining increased facilities for transportation of produce and merchandise between tide-water and the Western and Northwestern States; And whereas, it is desirable that such Convention should be fully and accurately informed as to the position of the State of New York in this respect, of the nature and extent of her public works, and of the great exertions and expenditures which her people have made in the construction of such works, as well as of their desire to co-operate with their Western brethren in the largest spirit of liberality consistent with justice, and the preservation of the public faith;

Therefore, be it Resolved, That the Auditor be requested on behalf of the Canal Board, to attend the said Convention, to the end that it may, so far as practicable, avail itself of his long experience, his accurate statistical knowledge in regard to the extent of our public works, their cost, the debt and taxation incurred in their construction, their capacity, the extent of their business and revenue, the rates of toll, and such other details as may be deemed important by the said Convention.

NOTE—AS TO THE PHYSICAL CHARACTER OF THE MISSISSIPPI.

The following information respecting the regimen of the Upper Mississippi and its tributaries, and which has a direct bearing on the nature of the water-communication between the Great Valley and the Great Lakes, is compiled from the "*Report upon the Physics and Hydraulics of the Mississippi River, by Capt. A. A. Humphreys, and Lieut. H. L. Abbot, of the Corps of Topographical Engineers U. S. A.; and published under the authority of the War Department,* 1861;—one of the ablest and most philosophical works, on the hydraulics of running water, to be found in the whole range of science.

MISSOURI BASIN.

This is much the largest of any of the tributary basins of the Mississippi, and has its sources in a region traversed by lofty mountain chains. The size of the river is disproportionately small compared with the area traversed. Its annual discharge is only about three-fourths that of the Ohio, although its basin is nearly two and a half times as large.

Its range between low and high water at its mouth, is about 35 feet; 20 at St. Joseph's, and at Fort Benton, about 6 feet. Its average high-water width for 600 miles above its mouth is 300 feet. The regimen of the Missouri above Milk river, according to Lieut. Grover, is as follows: Owing to the great elevation of its sources, the melting of the snows begins to swell the torrent early in the spring and goes on gradually to higher elevations, as the season advances, constantly diminishing until August, when it commences falling. The depth of water for the first of June is 3 feet; first of July, 2½ feet; first of August, 2 feet; and first of September, 1 foot.

The navigation of the Lower Missouri, according to Lieut. Warren, is generally closed by ice at Sioux city by the 10th of November. The rainy season commences between the 15th of May and the 30th of June, and lasts about two months, during which the river is in good boating stage.

The floods from melting snows reach the lower river about the 1st of July, which generally last a month.

The American Fur Company's boats are the largest class of freight-boats that navigate the Missouri, carrying from 150 to 200 tons to the Yellowstone, a distance of 1900 miles, and drawing from 3 to 4½ feet water. One of the greatest obstructions to the navigation of this river, consists in the great number of snags or trees, which render it necessary for the boat to lie by at night, and thus occasions a loss of nearly half of their running time.

The least low-water depth on the bars at Ft. Benton, Sioux City, St. Joseph and at its mouth, is 1.0 foot. Its width at Ft. Benton is 4,500 feet, and at the mouth, 3,000; the mean discharge per second is 120,000 cubic feet.

THE UPPER MISSISSIPPI RIVER

Is navigable 80 miles above the Falls of St. Anthony; but at this point occurs a perpendicular fall of 40 feet, with a formidable rapid above and below—the entire fall, being 65 feet. The range between high and low-water level is about 20 feet near Sandy Lake river; 20 at St. Paul; 10 at La Crosse; 12 at Prairie du Chien; 16 at Rock Island; 20 at Hannibal, and 35 at the mouth.

Its average width below Ft. Snelling is about one mile. The least low-water depth on the bar is 2.0 feet between St. Paul and Prairie du Chien, and 2.0 between Rock Island rapids and the mouth of the Missouri, and 2.0 between St. Louis and Cairo. The mean discharge per second is 105,000 cubic feet.

With regard to the Lower Mississippi, the least low-water, between Cairo and Memphis, is 5.0 feet; and between Gaines' and Red-river landing, 6.0 feet.

The width between the banks varies from 2,500 to 4,500 feet; the annual discharge is 21,300,000,000,000 cubic feet, or 75,000 cubic feet per second.

OHIO BASIN.

The Ohio river, throughout its whole length (975 miles) in low water, is a succession of pools and ripples, and is devoid of falls, except at Louisville, where it descends twenty-six feet in three miles. The range between extreme low and extreme high water is about 45 feet. At Wheeling it is 45 feet; at Louisville, 42 feet on the Falls, and 64 feet below them; at Evansville, 40 feet; at Paducah, 51 feet; at the mouth of the river, 51 feet. The least low-water depth on the bars, from the mouth of the river to Paducah, is about 3.0 feet; thence to Louisville, 1.5; thence to Cincinnati, 2.0 to 2.5 feet; thence to Wheeling, 1.0 foot.

The mean width of the river, between Pittsburgh and Pt. Pleasant is 1,000 feet at low water, and 1,200 feet at high water; but at the mouth it expands to about 2,500 and 3,000 feet. It discharges annually about five trillions of cubic feet, or about one fourth of the annual discharge of the Mississippi. The usual succession of the stages of the water is as follows: January, the river is frozen; February, breaking up, and high; March, high; April, high; May, falls somewhat; June, rises again; July, falls, and is low; August, very low; September, very low; October, very low; November, rises; December, well up. In August and September it is only navigable for boats of 18 inches draft.

It is evident, from these statistics, that a canal between the Valley of the Mississippi and the Great Lakes, preserving a uniform depth of seven feet during the period of navigation, from April to November, would afford a better navigation than that of any of the tributaries of the Mississippi, or even the main river itself. In the summer and autumn, while the river-navigation would be impeded, and, at certain points, practically suspended, the canal-navigation would be in perfection; on the other hand, in the winter and spring, while the canal-navigation was suspended by ice, the river-navigation would be at its highest stage. Thus, a choice of markets would be opened to every producer in the Great Valley, by a water-communication, cheap and expeditious, through which his crops would move as soon as harvested.

CORRESPONDENCE.

FROM HON. EDWARD BATES,

Attorney General of the United States.

WASHINGTON, May 27, 1863.

MY DEAR SIR:—Until two days ago, I had confidently hoped to be with you at the Canal Convention, to be held at Chicago on the second of June; but I am disappointed. And it is a real disappointment to me, for I am not only deprived of the pleasure and instruction anticipated at the Convention, but I am balked, also, in my intended visit to my own city, St. Louis, which I have not seen for more than two years.

It is my misfortune, and I am a great loser by it. But I cannot control the circumstances which seem to make it my duty to remain, for the present time.

I did feel an ambition to take some little part in the effort now in progress, to advance your great and beneficent object—to cast in one drop to swell the current of your noble enterprise. I am identified with the Great Valley by nearly fifty years of residence and labor in it; and I have some appreciation of its boundless capabilities, which need nothing to secure a success that will astonish the world, but the wise exercise of the good sense of the nation, and the prudent application of its means.

Our Valley presents an anomaly in geographical science. It is the only instance that I know of, where two mighty rivers occupy one and the same valley—and that valley the largest, richest, most inviting to human labor, on the face of the globe; for there is no mountain, no dividing ridge, between the Mississippi and the St. Lawrence.

I am really sorry I cannot be with you; for (to confess my weakness) I did wish to make a speech before your Convention, upon some of the topics which must come under discussion there—topics somewhat familiar to me for nearly half a century, and always full of interest and hope.

I remain, sir, with great respect,

HON. I. N. ARNOLD. EDWD. BATES.

FROM HON. CHARLES SUMNER,

A Senator of the United States, from Massachusetts.

WASHINGTON, May 27, 1863.

GENTLEMEN:—I must resign reluctantly the opportunity with which I am favored by your invitation, and content myself with reading the report of your powerful and well organized meeting at Chicago, without taking any part in it.

The proposition to unite the greatest navigable river of the world with the greatest inland sea, is characteristic of the West; the river is worthy of the fountain; and the fountain is worthy of the river. The mere idea of joining these together, strikes the imagination as something original. But the highest beauty is in utility, which will not be wanting here. With this union, the Gulf of Mexico will be joined to the Gulf of St. Lawrence, and the whole continent, from Northern cold to Southern heat, traversed by one generous flood, bearing upon its bosom untold commerce.

It will be for the West to consider well the conditions of this enterprise, and the advantages it will secure. Let its practicability be demonstrated, and the country will command it to be done, as it has already commanded the opening of the Mississippi. Triumphant over the wickedness of an accursed rebellion, here will be another triumph, which will be among the victories of Peace.

To this magnificent work science must contribute her resources. But there is something which is needed, even to quicken and inspire science; it is the unconquerable will, which does not yield to difficulties, but presses forward to overcome them. There is no word which is used with more levity than the word "impossible." A scientific professor declared in a public address, that the navigation of the Atlantic by steam was "impossible." Within a few weeks it was done. The British Prime Minister declared in Parliament that the construction of a canal between the Mediterranean and the Red sea was "impossible." The Pacha of Egypt, with French engineers, is now doing it. Mirabeau was right when he protested against the use of this word, as simply stupidity. But I doubt if this will be found in any Western Dictionary.

Believe me, gentlemen, with much respect,

Very faithfully yours,

CHARLES SUMNER.

To Hon. James Robb, I. N. Arnold, and others of the Committee, etc., etc.

FROM HIS EXCELLENCY, RICHARD YATES,

Governor of Illinois.

Executive Department,

Springfield, June 1st, 1863.

Gentlemen:—I received your invitation to be present at the Ship-Canal Convention, on the 2nd day of June, and though I hope to reach your city, yet my health will not be sufficiently recovered to take any active participation in the proceedings of the Convention.

I beg leave, however, most heartily and most fully to indorse the object of your assembling, not only as of most vital importance to Illinois, within whose boundaries one of the most important works to be considered by you, is to be located, but as affecting the interests, welfare and destinies of every State so materially, as to challenge the hearty and united co-operation of the whole country.

I regard this enterprise as combining, in three comprehensive degrees, principles and purposes which should commend it to every patriot and philanthropist.

That which adds to our commercial prosperity, increases our military strength, and tends to cement our Union, should certainly enlist a large share of our earnest consideration.

Deeply solicitous as to the result of the great struggle, now going on for our glorious Union, (in which, may God grant the victory to our noble patriots perilling their lives upon the battle-field for the cause of Union, liberty and morality,) I look upon the completion of this truly national enterprise as next in importance to the salvation of the country.

He who establishes his claim as a patron of the work, helps to advance in the highest degree our national prosperity, to establish our military supremacy on the continent, and to bind in more indissoluble bonds the States of our Union. He proposes by this most important link in the chain of international communication, to enable the General Government to grasp the great Father of Rivers in one hand, and the great system of Northern Lakes in the other, and to say to all enemies of our Union:

"What God and Nature, the arts, sciences, and civilization have joined together, let not the spirit of disunion, the incidental obstacles of material progress, or the barbarism of a society founded upon a system of unrequited labor, put asunder."

The completion of this work would bring our agricultural and manufacturing interests into closer proximity, and would add immensely to the amount and profit of our productions; while by the saving in transportation, the consumer in our own and foreign countries would receive every commodity at a much reduced price. Already the *London Times* and other organs predict that, if this and similar works connecting the Western States with the seaboard are completed, the farmers of Great Britain must turn their attention to other crops than cereals; while the Irish newspapers assert, that the agricultural products of that country, for which Great Britain has heretofore been the market, could not withstand the competition with this country.

Under the present incomplete means of transportation, it costs more than the price at which corn may be profitably grown in Illinois, to convey it to the seaboard and to Europe; and it is only in time of famine or great scarcity that we are able to compete with the dependencies of Great Britain in her markets. Let, however, enlarged facilities of transit be created between the West and the seaboard, and Indian corn may be used in Great Britain, not only as it now is, for human food, but for fattening cattle, hogs, sheep, etc. And when these facilities are created, which will be very soon, if Eastern and Western enterprise unite in the great purpose, the Illinois farmers, by the aid of improved agricultural and improved farming machinery, can distance all continental European competitors in the British markets, while the agricultural supremacy of the Western States will be established for all time in the world's markets.

With soils of exhaustless capacity for production, with a hardy and intelligent yeomanry, and with such rivers, railroads, canals and lakes, Agriculture will have her millenium in the North-West, and reap harvests such as the world never saw; and at the same time, by day and by night, myriads of engines will bring back, over the same channels of trade, the manufactured products of the East, to supply an ever-increasing demand, thus adding immensely to the profits of every producer, manufacturer and laborer, and to the general wealth of the country.

As a question of military supremacy, the merest tyro in the art of war must see that the power which possesses the naval command of the Lakes must have the command of all the territory bordering on them, either for offense or defense. The Mississippi pierces the Cotton States of the South, divides them in twain, and gives them virtually into the hands of the power that has the control of that stream. The Lakes of the North pierce the great agricultural region of the North-West, and give it into the hands of the power that controls them. We cannot, in the event of a war with England, successfully carry on military operations against Canada, unless we hold the Lakes as the basis of such operations. Nor can Canada successfully engage in military operations against the Western States, unless she controls the Lakes as the basis of such operations.

It is related that, towards the close of the last war with Great Britain, the ministry of that country proposed to the Duke of Wellington that he would take command of the British army in Canada. That able general and statesman, it is said, answered that he could only do so on condition that his land operations should be seconded by a fleet upon the Lakes, sufficiently numerous and powerful to sweep them throughout their whole extent. This not being conceded to him, and probably rendered impossible by the victories of Perry, McDonough, and others of our naval heroes of that day, or peace having shortly afterwards been declared, the project of sending the conqueror of Napoleon to head the armies then operating against the United States was abandoned. But the British Government never has forgotten the principle upon which the Duke of Wellington proposed to conduct the campaign. It immediately proceeded to complete a series of canals in the Canadas, which have not only added greatly to the commercial prosperity of these Provinces, but also rendered it possible for the mother country, at any time, to undertake offensive operations against the United States, without the necessity of maintaining an additional fleet upon the Lakes.

At present, then, the power to control the great Northern Lakes is in the hands of the British Government, since by her system of canals, she can, in a few days or weeks at farthest, place a fleet of two hundred or two hundred and fifty gun-boats (always kept in readiness in her home dock-yards) upon them. This fleet could not only annihilate our commerce, but place all of the lake cities, without a single exception, at the mercy of the invader. We would thus commence the war under disadvantages, which only the most heroic exertions and most terrible sacrifice of men and material could overcome.

Give us, however, a water-communication of the seaboard and the Lakes with the Mississippi and its tributaries, by a ship-canal of superior or equal dimensions with those of the Canadas, and we would be placed on equal terms with Great Britain, and readily transfer our fleet of gun-boats and transports and munitions of war to the Lakes, and our armies to their shores, ready to drive back the invading enemy.

But, if it is granted that such a work would add greatly to our military power and naval prestige, and commercial facilities, then its benefits as an additional bond of Union, must be admitted. It would indeed be an additional link to the chain of artificial communication which now binds the East and West, and the first of which was forged, in so many and appaling difficulties, by the ever-to-be-remembered De Witt Clinton, when he completed the New York and Erie canal. That work was the first and enduring triumph of art over nature, on the continent. It has been followed by many other victories and achievements of American genius and enterprise, scarcely less renowned than those of war itself. We have but to look at the vast net-work of railroads, traversing every part of the country; the long trains of cars which sweep at the rate of forty miles per hour along the cragged sides of the Alleghanies; the untiring tramp of the iron horse over unmeasured leagues, across rapid rivers, and through perforated rock, and cliff, and hill; to be convinced that there is no undertaking, within the range of possible achievement, which cannot be accomplished by American genius, skill, and enterprise.

The time has now come, and will not longer be delayed, when the last and strongest connecting links in the chain, and by far the most important works in a military and commercial point of view, must be completed. We cannot look over the magnificent outlines of this country without seeing that the works proposed by this Convention are not only material and proper, but in the course of events inevitable. The Mississippi and Missouri, 4,000 miles in length, and with their tributaries, comprising 20,000 miles of inland navigation, with numerous flourishing States and Territories upon their banks, and floating hundreds of millions of commerce on their currents, have for some months been closed to our commerce, by a blockade which could not have been maintained for one month, if the ship-canal from the Illinois river to Chicago, and uninterrupted water transit to the Eastern cities, had existed. The existence of actual war is an argument which our people fondly hoped never to have had, but it now loudly demands the construction of this canal as the first military necessity; to prepare the country not only for the highest commercial advancement, but for any warlike emergencies which may arise, and as a perpetual bond of union between the Eastern and Western and Gulf States. Railroads penetrating our land throughout its whole extent, our rivers and lakes improved and united, ocean brought to ocean, the East to the West and the South to both, the strong cords of social and commercial intercourse will be stronger bonds of Union than all the constitutions man ever framed. Rome maintained her supremacy by means of her public highways. They extended from the city a distance of four thousand Roman miles. They reached to her remotest frontier, established her supremacy, and floated her imperial eagle over one hundred and twenty millions of subject people. They united to herself, as to a common centre, all the distant parts of universal Empire.

In the consideration of these great measures, there is no propriety in any appeal to sectional prejudice; for, if heretofore the construction of a ship-canal from the Illinois river to Lake Michigan may have been considered not as a national work, and therefore not a proper subject for appropriations from the Federal Treasury, that objection no longer exists.

I would not say a word in disparagement of the patriotism and valor of the Eastern or Middle States, for in every battle of the present war they have fought with more than Roman heroism, and not only in victory, but in the face of defeat and fearful slaughter, stormed the most impregnable batteries, never giving an inch or yielding ground till the order to retreat was given. At the same time, all will admit that the North-West has also come up nobly to the support of the war and the Government. Covered all over with glory, resplendent with the lustre of the achievements of her sons on so many glorious battle-fields, and still ready to pour out the last drop of her blood, and to exhaust the last cent of her treasure to restore the Union and to save from destruction of traitor hands this last hope of humanity, and this beautiful temple of human freedom, she now trusts and believes, and this Convention is assurance of the fact, that in the disbursements of Federal money for military and naval purposes, the North-West will not be denied the small sum for the object this Convention has in view, essential in time of war for her defense, and to the protection of the whole Union.

Thank God, I have no fear of disunion. God himself has written Union upon the face of our country. Its lakes and rivers, and railroads, its whole configuration, proclaim it the home of one people—and that people increasing in numbers and moving on in their mighty career, renowned in arts, in science, in arms, and presenting to mankind the example of a government, happier, more renowned, glorious and free than the history of the world has yet recorded. Upon her past history mighty memories rise and cluster, but they are as nothing to the visions of splendor, power and grandeur, which under enterprises like this, for which you are assembled, are to blazon the pathway of her glory in the future.

From early boyhood to the present hour, I have had the bright dream that God had showered down his blessings upon this my native land, as the chosen theatre upon which civilization, the arts and sciences, Christianity and freedom, should display to mankind the loftiest achievements of a true Christian civilization, and the proudest national glories; and, by the blessing of God, I will continue to believe, through all the years of the future, as long as the Mississippi shall flow, or Lake Michigan shall heave with surging billows, that from the Atlantic to the Pacific, from the Bay of Fundy in the North to the Gulf of Mexico in the South, we shall have one unbroken Union of States, whose hundreds of millions of people shall speak the same language, live under the same constitution, and high over all shall float the same flag of freedom and National Union.

In conclusion, trusting that a spirit of harmony may attend all your deliberations, and that they may conduce in the highest degree to the noble and useful objects at which they aim,

I am, gentlemen, your obedient servant,

RICHARD YATES.

To James Robb, Esq., Hon. I. N. Arnold, and others, Committee.

FROM HON. JUSTIN F. MORRILL,

A Member of the House of Representatives from Vermont.

Stafford, Vt., May 25, 1863.

Dear Sir:—Until recently I have expected to be present, by your invitation, at the Ship-Canal Convention, but at this moment, though still proposing a Western journey, I find some business engagements will retard my departure, so that I may not reach Chicago until some days after the 2nd proximo.

The importance of a great canal—greater than we yet possess—between the Mississippi and the Lakes, will hardly be questioned in any quarter, and is obviously on a par with that connecting the Lakes with the Atlantic. It is a link in the magnificent American chain of water-communications—coast, lake, river, and canal—circling more than half the States, and draining—therefore, hardly less useful—most of the remainder, and its perfection will not be and ought not to be abandoned, nor long postponed. Our annual experience shows the existing means of water-transit across our northern borders are inadequate. The recent enlargement of the New York and Erie canal is such a financial success that it may be reasonably expected to assure a further and sufficient enlargement for any probable requirements. But the Illinois portion of the great business-channel exhibits the greatest inadequacy, involves the largest expenditure, and merits the earliest consideration.

Those in favor of developing the vast dormant land-capital of the Western States, of increasing the value of labor, and of unfolding and securing the power of the United States, have often eloquently portrayed the value of the contemplated improvements, and their arguments cannot be bettered by repetition.

The New York and Erie canal has not only been productive on the capital invested, but has created much wealth beyond its local limits—fertilizing the sources of its business quite as extensively as its *debouchure*, where it discharges its overflowing commodities, and inundates the great marts of the world.

The natural vent of the Lakes, the St. Lawrence, runs too near polar ice, as well as through a foreign land. Commerce and travel required a more direct and feasible outlet, and found it through the colossal enterprise of DeWitt Clinton. In the absence of this, many of the now thrifty denizens of the West would have been too poor to have reached their homes in the New World, or would have been lost in Canadian forests; and they must, therefore, have remained in other hands, or hung like drift-wood on the shores wherever landed.

The utterance of truth only reveals the magnitude of such works. The still small waters, under the guidance of skillful engineers, are potent in turning the business of the world into its most stirring and lucrative channels.

A glance at the map discloses the fact that the proposed ship-canal, compared with the New York canal, and with an equally unfailing capacity, will attract the commerce of a territory not inferior in extent nor less abundant in its productions. No more need be asked. These productions, though now mainly pastoral and agricultural, may be expected, at a very early day, through the aid exuberant coal has supplied for cheap steam power, to assume all the forms of diversified manufactures, and thus concentrate, perhaps, a denser population than will then be found elsewhere upon the American continent.

The work is bound to succeed. If only private enterprise embarks the capital required, it will go on to substantial completion, and being thus most economically managed, will take rank among the solid dividend-paying stocks.

The work is bound to succeed. If the State of Illinois alone should take charge of the work, who would doubt it? Her resources are ample, and the ultimate return with gain is nearly if not quite absolute.

The work is bound to succeed. For the day will soon come, I am persuaded, when the Federal Government, without peril to its own vitality, will be able to devote some share of its large resources to an object of conceded national importance. If the work to be done is great, the want of it is equally great and ever increasing. Even had the passage to be hewn out of the rock, I should look for its accomplishment. It is but the removal of earth, and requires the spade.

I have often seen portions of the prairies of the West, and always with increasing admiration. Large numbers of people have removed thither from this section of the country, and like other Eastern points, transferred to a more fertile soil, I know their dimensions always expand. I shall ever hail with joy whatever contributes to the the improvement of any portion of our country or its people, and especially of the West, "for we be brethren."

I regret that I cannot be present on an occasion of so much interest.

Very truly yours,

JUSTIN S. MORRILL.

Messrs. JAMES ROBB, I. N. ARNOLD, and others, Committee, Chicago, Ill.

FROM HON. W. A. RICHARDSON,

A Senator of the United States from Illinois.

QUINCY, ILL., May 16, 1863.

MY DEAR SIR: I am in receipt of yours of the 14th. I fear it will be impossible for me to be at Chicago on the 2nd of June. Mrs. Richardson is very ill, so much so that I cannot leave her bed-side only for a few hours at a time. While I hope for her restoration to health, I cannot hope such improvement as will enable me to leave her by that time. I will only add that I regard the construction of a ship-canal from the Lakes to the Mississippi river, as equal in importance to the whole country, either in war or peace, to any that has ever been proposed on the continent. I regret that I cannot hope to be with you at the time proposed.

I am truly,

W. A. RICHARDSON.

FROM HON. S. C. FESSENDEN,

A Representative in Congress from Maine.

ROCKLAND, ME., May 20, 1863.

Messrs. JAMES ROBB, Esq., Hon. I. N. ARNOLD, and gentlemen of the Committee on Invitations:

It is with pleasure I acknowledge the receipt of your note of April 16th, inviting me to be present at a National Convention to be held in Chicago on the 2nd day of June next, for the purpose of considering the importance of enlarging the canals between the Mississippi and the Atlantic, with the view of increasing their efficiency as national, commercial and military works, and as tending to promote the development, prosperity and unity of the whole country.

I regret my inability, arising from business engagements, to participate in the proceedings of the Convention.

May I be permitted to add—the Convention cannot even estimate the importance, in my opinion, of the object which it has in view. During the session of the 37th Congress, of which I had the honor of being a member, I gave this subject no little attention; and, as the results of my investigations, I was convinced, that such was its relation to the material interests of the nation, in any point of view, that, could the people but have the light which already it is in the power of facts to give them, they would be absolutely importunate, until the object is accomplished. Conseqently the "Act to construct a ship-canal for the passage of armed and naval vessels from the Mississippi river to Lake Michigan, and for the enlargement of the locks of the Erie canal and the Oswego canal of New York, to adapt them to the defense of the Northern Lakes," had my early and most earnest support; and no man more deeply regretted its failure to pass the House than myself.

It is my hope, therefore, that the Convention will lay *the facts before the people*, and I predict such measures, as the result of this course, as will speedily accomplish the work—a work which will so greatly augment the prosperity, and contribute to the unity of the country, that its projectors will be remembered amongst their country's benefactors.

I am, gentlemen, very respectfully,

Your ob't serv't,

SAM'L C. FESSENDEN.

FROM HON. J. M. EDMUNDS,

Commissioner of the General Land Office.

DETROIT, MICHIGAN, May 26, 1863.

Hon. I. N. ARNOLD. Sir: I have to thank you for the circular of invitation to the National Canal Enlargement Convention, to be held at Chicago on the 2nd of June next. Prior engagements are likely to prevent my compliance with your request; but neither the presence nor absence of any individual can materially affect the onward progress of the great measure you have under consideration.

The military, commercial, and agricultural necessities, and the interests of the country, alike demand the speedy completion of the contemplated improvement. When the Pacific railroad shall have stretched its iron ligaments from the Mississippi to the Sacramento, and the flood of commerce and travel which must seek this route from Eastern Asia and Western Europe, shall meet in the Valley of the Mississippi, the vast products and population of that region, the present facilities of movement will be literally blocked up—suffocated, so to speak.

The great Lakes and the Mississippi combined, can alone afford relief, and their waters must be made to commingle by a channel that can bear, without hindrance, the naval and commercial marine of the country, and the moving masses of people residing upon their shores and borders. And it is of equal importance that these facilities be extended to the Atlantic, within the territorial limits of the great Republic.

When the power of this nation shall be vindicated, and its perpetuity secured, by the complete suppression of the existing infamous rebellion, as soon will be the case, we shall as a people enter upon a career of enterprise, prosperity, and grandeur, never before witnessed. Wealth, population, and power will flow in upon us from every point of the compass. Our railroads, rivers, lakes, mines, plains, and forests, all will be made to contribute to our greatness, and facilities for the movement of property and people will be demanded faster than they can be constructed. San Francisco, St. Louis, New Orleans, Pittsburgh, Chicago, Detroit, Buffalo, New York, and intermediate localities, will become the points of distribution for the congregated wealth and population of the country, and will grow with its developments.

They must be connected by ample facilities of transportation and travel. I have neither time nor space to allude to the economical advantages of your projected improvement. These are apparent, and may be proved to the satisfaction of all. You must persevere, and you will certainly achieve success, to which I will gladly contribute, to the extent of my power and ability.

With great respect, your obedient servant,

J. M. EDMUNDS.

FROM HON. D. DAVIS,

One of the Justices of the Supreme Court of the United States.

INDIANAPOLIS, May 27, 1863.

Messrs. JAMES ROBB, I. N. ARNOLD, F. C. SHERMAN, and others, Committee on Invitations:

GENTLEMEN:—I regret exceedingly that my judicial duties will deprive me of the pleasure of meeting with you on next Tuesday. I feel a deep solicitude for the success of the Convention. It cannot fail to enlighten the public mind; and, if it shall be the means of awakening a general interest on the subject for which it is convened, it will have accomplished a great result.

With high respect, yours most truly,

D. DAVIS.

FROM COL. S. H. LONG,

Chief of the Bureau of U. S. Topographical Engineers, who, for many years, was personally in charge of the Government Surveys of the Western Rivers.

WASHINGTON, May 20, 1863.

To Gen. JOS. G. TOTTEN, and others, delegates to the Chicago Canal Convention:

GENTLEMEN:—Finding it very inconvenient, and even quite impracticable, to meet you as a member of the Convention, I beg leave to submit my views in relation to the contemplated ship-canal, which may throw useful light on the topics of inquiry likely to occupy your attention.

In the month of October, 1816, I first visited Chicago, by a route from St. Louis, leading through a savage and roadless wilderness, *via* Fort Clark and the valley of the Illinois river, to Lake Michigan.

At that time, the Chicago river discharged itself into the lake over a bar of sand and gravel, in a rippling stream, ten to fifteen yards wide, and only a few inches deep.

About ten miles southerly of Chicago, the Little Calumet entered the lake, over a similar bar, but in a broader, deeper, and much more copious stream.

Near the southerly extremity of the lake, the Grand Calumet enters the latter; but at the time of my first visit, its mouth was effectually blocked up by a high and dry sand-bar.

It will hereafter be shown, that the waters of both Calumets are sometimes discharged into the lake through the mouth of one, and at other times through the mouth of the other. It is also said that, in wet seasons, a part of their waters is discharged into the Des-Plaines river.

In June, 1823, I again visited the localities just mentioned; the country still destitute of permanent settlement or improvements. At this time the discharge from the Chicago river was similar to what it was before, while the discharge from the two Calumets was materially changed; the mouth of the Little Calumet being blocked up by a dry sand-bar, and the discharge of both being through the mouth of the Grand Calumet.

In the same year and month, I ascended in a perogue, by way of Chicago river and its southerly branch, to Mud Lake, which discharges itself in two directions; in one direction into Lake Michigan, and in the other, into the river Des-Plaines.

In connection with the subject of the proposed ship-canal, there are certain considerations worthy of particular attention. At a short distance inland from the mouth of the Little Calumet, are two considerable ponds or lakes, said to be deep and navigable. These may be converted into basins or ports, rendered accessible by suitable connections with the proposed canal, and affording ample accommodations for shipping. Moreover, the Little Calumet is probably quite as deep and navigable as the Chicago river, and to an extent considerably greater than that of the latter.

The broad and extensive plain of the Chicago (Che-Kan-Ko, *Wild Onion*), is altogether diluvial, and is generally diversified by flat prairie, sand-ridges covered with a woodland growth, swamps, slashes, sluggish streams, stagnant pools, etc. The sand-ridges, hills, and knobs, rise from a few feet to ten, twenty, and in some instances thirty feet above the surface of the plain, and are generally covered with a stinted growth of oaks, bushes, vines, and furze.

Besides the streams already mentioned, the plain is intersected by numerous others of less magnitude, and with equally sluggish currents, in all possible directions.

The entire plain is no doubt underlaid by an extensive bed of solid rock (limestone and sandstone), at depths of only a few feet below its common surface.

At the distance of about thirty-four miles from the lake-shore, the Des-Plaines assumes a more lively current, and descends by gentle slopes to its junction with the Kankakee, or head of the Illinois river. From this point, the Illinois descends with greater rapidity, passing occasional rapids of greater or less extent, till it reaches LaSalle or Peru, where it becomes a sluggish stream, and continues so quite to its mouth.

In September, 1816, I ascended the Illinois to the head of Lake Peoria in a small keel-boat, and literally passed through extensive fields of wild rice, springing from the river bed and rising virtually several feet above the water-surface, the current being too sluggish to sway down the straws. The river continued in a similar condition till the frequent passage of steamboats prevented the upward growth of the rice.

It is deemed pertinent in this place, to introduce a variety of results derived from the surveys for the Michigan and Illinois canal, made in 1830–31, under the direction of the late Dr. William Howard, then Civil Engineer in the employ of the Topographical Bureau.

The survey was commenced at a point on the southerly branch of the Chicago river, about five miles from the lake-shore by the meanderings of the stream. Running thence in a southerly direction, and a few feet only (3 to 10 feet) above the surface of Lake Michigan, it entered the river Des-Plaines at a distance of twenty-nine miles from the place of beginning, or thirty-four miles from the lake-shore. The greatest elevation of ground above the surface of the lake, on the line surveyed, was only about fourteen feet, and that only at a single point; indeed, the general surface of the spacious plain has an average height of only about ten feet above the lake.

The rock stratum underlying the Chicago plain, has a depth of three to eight or ten feet below the surface of the ground, as determined by frequent borings along the line surveyed.

The survey of a line about eighteen miles long, for a feeder, intended to connect the Little Calumet with the main canal in the valley of the Des-Plaines, indicates that the surface of the ground traversed by it, has about the same elevation as that traversed by the canal-line.

At the distance of thirty-four miles from the lake, the surface of the Des-Plaines becomes co-incident with that of the lake, and at the point just indicated, the Des-Plaines begins to fall below the lake-level, and continues to descend thence to its junction with the Kankakee, and thence with the Illinois river to La Salle, or the lower terminus of the present canal. The distance from the point mentioned, to La Salle, is sixty-seven miles, and the aggregate fall in this distance is 137½ feet, or a little more than two feet per mile, making the entire distance from the lake-shore to La Salle, about 101 miles, and the aggregate fall below the lake-surface, 137½ feet as before.

From La Salle, downward to the mouth of the Illinois, the distance is computed at 220 miles; and throughout this distance the river current is remarkably sluggish, the aggregate fall being represented as twenty-six feet only, or, on an average, less than one and one-half inches per mile. Hence, it is obvious that a comparatively moderate supply of water from the lake in low stages, will contribute to render this portion of the river quite as navigable as the Mississippi, from the mouth of the Illinois to Cairo, at the mouth of the Ohio, in similar stages.

In that portion of the Illinois last considered, there are occasional hard bars, across which the channels are narrow and crooked, and will require enlargement and straightening. The channels across the low-water bars of the Mississippi will also require to be made wider and deeper. This may be effected, in both cases, by the use of scrapers, similar to those successfully employed in opening and keeping open the channel across the formidable and indurated bar at the South-west Pass of the Mississippi.

I now conclude my remarks with a few observations and suggestions in reference to an open and through-cut ship-canal, from Lake Michigan to the Des-Plaines river.

1. The elevation of the summit-pool of the present canal, is eight feet above the surface of Lake Michigan.

With regard to its enlargement to the dimensions required for a ship-canal, it is proper to observe, that its trade is now and has been for several years past, of too much value and importance to justify its suspension, during the period required for the construction of the contemplated through-cut canal. Moreover, its enlargement would render unavoidable the destruction of a vast amount of valuable property, situated on both sides of the canal.

2. One of the main objects in view, is the conversion of Lake Michigan (Miche-Sagahegan, *Big Lake*), into a grand reservoir, to supply not only the canal, but the Illinois river, with water enough to render them both navigable at all times when not obstructed by ice. Of course, the summit-pool must extend from the lake, through a distance about equal to that from the lake to Lockport, viz., about thirty-four miles.

The depth of water in the ship-canal should be at least eight feet; its width one hundred feet, with passing places one hundred and fifty feet wide, and one mile apart. The declivity of the canal-bottom should be about two inches per mile, from the lake to the Des-Plaines river.

3. The lock-chamber at the westerly end of the pool, and at all other points below, should have a width of about 60 feet; a length of 220 feet, and a lift varying from four to ten or twelve feet. The upper lock should be accompanied by spacious waste-wiers or gates, perhaps both, sufficiently large to supply the requisite lockage-water, and an additional volume of surplus water sufficient to render the Illinois navigable from La Salle to the Mississippi.

4. The sides of the canal, especially of the summit-pool, should be revetted with substantial stone-walls, resting on the surface

of the rocky substratum of the plain, which rises from beneath the lake-surface, to the height of nine or ten feet above the same, and through which the prism of the canal must be excavated; the excavations thus formed, will afford ample materials for the walls.

5. In connection with the summit-pool, there should be one or more spacious inland basins or ports, large enough to accommodate the shipping employed in the lake-trade, and for various other purposes.

6. An artificial harbor, like that at Chicago, may be formed at almost any point on the southerly coast of the lake, and an entrance to it may be opened across the lake beach, by sinking parallel piers and removing the sand and gravel from between them by dredging. This mode of forming an entrance, as I believe, was first proposed by me in 1816.

In view of the rapid growth, and rapidly increasing trade of Chicago, there can be no doubt that in the lapse of a few years, one or more additional harbors, for the accommodation of its commercial transactions will be required, at Chicago.

7. It is believed that a route quite as eligible for a canal, as that pursued by the present canal, and passing over ground equally as favorable, if not more so, may be discovered and adopted as the route for the proposed ship-canal.

8. With regard to the proposed depth of the canal, it is undoubtedly true that no vessel or other craft, drawing more than six feet of water, can navigate the Mississippi, between the mouth of the Illinois and Cairo, more than six months in each year. Hence, although a greater depth may seem desirable, on many accounts, yet, in this case, it would be likely to prove useless.

9. Whatever craft is employed in navigating the canal, it should be driven by propellers, of suitable construction, in order to prevent abrasions on the earthy sides of the canal.

10. The lake being regarded as a reservoir, the canal will serve not only as a channel of trade, but as a feeder to supply both the canal, and the Illinois below the canal, with the requisite quantity of water to render both navigable at all times. Hence, the waste-wiers at the locks, and especially at that connected with the summit-pool, should be constructed in such a manner that the quantity of water passed by them may be increased or diminished, according to the exigences of the river below the canal.

11. The selection of a route most commodious and favorable in all respects should be preceded by, and based upon careful, thorough and judicious surveys, executed by experienced and skillful engineers; economy of construction as well as facility and dispatch of transportation, being the leading objects kept in view.

Respectfully, gentlemen, your obt. servt.,

S. H. LONG, Col. U. S. Engrs.

FROM HON. EDWARD EVERETT,

Late a Senator of the United States, and Minister Plenipotentiary to Great Britain.

BOSTON, 16th May, 1863.

GENTLEMEN:—I received, a short time since, a copy of your invitation to the National Convention, to be held at Chicago on the 2nd of June, to consider the importance of enlarging the canals between the Mississippi and the Atlantic. It will not be in my power to attend the proposed meeting, but it gives me great pleasure, in compliance with your request, to express the most favorable opinion of the enterprise in question.

I had the good fortune last year to go over the greater part of the ground traversed by the proposed enlarged line of communication, and to form, from personal observation, a distinct idea, both of its practicability and importance. I know of no public work, within the limits of the United States, at all comparable in importance with a communication like that proposed between the Atlantic and the Mississippi. The resources of the West are boundless, and still in the infancy of their development;—nature has formed the two great sections of the country, East and West, for the most intimate and mutually beneficial relations with each other. Much has already been done to open those pathways, to which the hand of the Almighty Engineer has pointed the way, and nothing is wanted to complete the work, but an appropriation of the resources of the country, which, with reference to the importance of the objects to be promoted, ought not, however considerable in itself, to occasion a moment's hesitation.

With cordial wishes for a harmonious and successful meeting,

I remain, gentlemen, your friend and fellow citizen,

EDWARD EVERETT.

Messrs. JAMES ROBB, J. W. SMITH, GEO. SCHNEIDER, and others, whose names are subscribed to the invitation.

FROM HON. J. M. HOWARD,

A Senator of the United States from the State of Michigan.

DETROIT, May 30, 1863.

MY DEAR SIR:—I much regret that in consequence of professional and other engagements, I shall be unable to attend the Convention which is to meet in Chicago on the 1st proximo, to promote the great cause of the enlargement of the canals between the Mississippi and the Atlantic. I beg, however, to assure you that no one can be more sensible of the great national importance of the project, or more alive to the benign policy of knitting together the East and the West by the ties of mutual interest. Before God and the world we have pledged ourselves to be not merely one people, but a united people; and unless we are blind to our own interests, and recreant to the principles, the promises and the fame of our ancestors, we shall keep that pledge good. I beg to assure the Convention, that whatever the constitution and the state of the country permit to be done in furtherance of this great object, will not fail to receive my earnest and cordial countenance and support, whether in a public or a private capacity.

I remain very truly and sincerely your friend and ob't serv't,

J. M. HOWARD.

Hon. I. N. ARNOLD, Chicago.

FROM J. A. HAMILTON, ESQ.,

The surviving Son of Alexander Hamilton.

NEVIS, DOBBS FERRY P. O., May 19, 1863.

To the Hon. JAMES ROBB and others, Committee on Invitations, Chicago:

GENTLEMEN—I received to-day your invitation to attend the Convention to be held at Chicago on the 2nd day of June next, which I accept with much pleasure.

I have a grateful recollection of the attentions I received in your city, when, a score of years ago, I attended a convention there on a kindred subject. At my advanced period of life it is presumptuous to say I will make so long a journey; but I will attempt it. Next to crushing out this wicked rebellion, nothing is so near my heart as this patriotic effort " to promote the development, prosperity and unity of our whole country."

I have the honor to be, with great respect, your ob't serv't,

JAMES A. HAMILTON.

FROM HON. T. O. HOWE,

A Senator of the United States, from Wisconsin.

GREEN BAY, May 29, 1863.

GENTLEMEN:—I have to acknowledge the receipt of your note of the 29th April, inviting me to attend a Convention, to be assembled at Chicago, on the 2nd of June next, to consider the project of enlarging the " canals between the Valley of the Mississippi and the Atlantic."

I have delayed reply to that note, up to this time, in hope of finding myself able to accept your invitation.

Now, to my regret, I am obliged to say, I shall not be able to visit Chicago upon that occasion.

Let not my enforced absence be thought to evidence any want of sympathy for the great objects which will engage your attention.

Whether I regard the military or the commercial aspects of the measures proposed, I hold them to be eminently national, and eminently sagacious.

It has been denied, that the enlargement of these channels has any importance in a military point of view. I fear such denials are made upon a very imperfect consideration of the subject.

The extent of our coast upon the Lakes is nearly equal to that upon the Atlantic. I affirm, that, to-day, it is of vastly more importance to the nation. Along the Atlantic coast are found the entrepots for our foreign imports. These, for the most part, consist of the luxuries of the nation—of articles which we ought to produce at home, or ought to go without.

Along the Lakes flows the *food* of the nation; without which it could not live. The lake-coast is as accessible to the fleets of one foreign power, and that the greatest, in a maratime point of view, as the Atlantic coast. To defend the Atlantic coast, and its commerce, our government expended, before this war commenced, on fortifications and on naval armament, not less than fifteen millions annually. To defend the lake-coast, and its vital commerce, we have not a single fortification worthy of the name; and we have not only no naval armament, but we are absolutely prohibited, by treaty stipulations, from maintaining upon the Lakes more than three vessels of one hundred tons each, armed with not more than one eighteen-pound cannon each.

Let any man consider the relations of this lake-commerce to the nation; what New York and New England would be, if they were cut off from this flow of food; what the West would be, if its currents were turned back upon itself; and he cannot fail to perceive that a permanent obstruction in this great artery would be as fatal to the vigor of the nation, as a cord around the neck would be, to the vigor of a man. The West would suffocate, the East would starve.

Before the war, New York thought she lived upon her participation in the annual export of two hundred millions of cotton. Cotton left her—left her two hundred millions in debt; and she has never missed either the money or the export.

A few months ago, New York trembled before the possibility of a visit from the Merrimac. She knows, to-day, that a few hostile gun-boats, placed before the harbor of Buffalo, would be almost as fatal to her as the whole British navy in her own bay.

It is said, and said truly, that we can abrogate the treaty in six months; whereas, it would require years to open our canals to the admission of boats suited to the purposes of naval defense. But the abrogation of the treaty would not provide for the emergency. The abrogation of the treaty would only restore the right to build a navy upon the Lakes, it would not build a navy there. To build a navy upon the Lakes, equal to the defense of this magnificent frontier, and to maintain it three years, would cost more than is asked to open them to the navy of the nation, and to whiten them with the commerce of the world.

The United States have only these three alternatives to choose between: either the splendid empire, washed by these interior seas, must remain utterly defenseless against the perils of foreign war, always possible; or, we must open the water-channels connecting them with the Mississippi, or with the Atlantic, to the admission of our naval fleet; or, we must give notice of our intention to do so, and in six months, commence to provide a separate navy upon the Lakes, to be used only in a brief possible war, and to be useless through a long probable peace.

The first alternative is too full of hazard to be adopted by a people really independent: the last is too costly to be adopted by a people truly economical.

Unity of commerce, unity of coast, and unity of defense, will secure the unity of the Republic, and its absolute independence.

But I am free to confess, that, in my own judgment, the commercial importance of the measures proposed is scarcely less than their military importance. As military measures, they will only subserve the wants of war. That, if we are just in the conduct of our foreign relations, and wise in the administration of our domestic affairs, we may expect will be the exception to our national life.

As commercial measures, they will largely minister to the wants of peace. That, after a few more months of patient endurance and heroic endeavor, we may hope will be the rule of our national life.

The limits of a letter will not permit me to discuss, or fairly to present the economical aspects of the measures proposed. But if the national growth were to be permanently arrested to-day, and the national production were never to exceed its present volume, and if the present product was exclusively agricultural, and was limited to the two staples of corn and wheat, it would, nevertheless, remain that one hundred million bushels of those cereals pass over the Lakes annually.

It is claimed that the expenditure of three millions of dollars will expedite and cheapen the movement of these products, to the extent of fifteen cents per bushel.

It is true, those commodities are mainly the product of the States termed North-western, and are mainly consumed by the States termed North-eastern, and by Europe. Hence, it is not unusual to hear it said, that the benefits to be derived from this expenditure, would be divided between the producers and the consumers of these products, and so would be sectional and not national in their influences.

But it should be remembered, that these commodities are, for the most part, consumed by the manufacturing industry of this and other countries. They are exchanged for their fabrics. They have a direct and important influence upon the value of those fabrics. And in the value of those fabrics, every village, and almost every family, in the Union, has a direct interest. The corn of Illinois, and the wheat of Wisconsin, it is true, never cross the Mississippi, in bulk; but when transmuted into the fabrics of Lowell and Birmingham, they cross the Rocky Mountains. Whatever lessens the prime cost of food, tends to lessen the prime cost of every article into which it enters. It is, I think, the *tendency* of McCormick's reaper, no less than of the Erie canal, to lessen the cost of cloth at Santa Fe. The effect, to be sure, is not so perceptible, but it is as actual.

Hence, it is presumed, but few statesmen will be found to deny, that if the expenditure of three millions of dollars, upon the canal between the Lakes and the Hudson, will send the wheat and the corn of the North-West, to their consumers, at an annual saving in the cost of transportation, of fifteen millions, it is an object of deep interest to the whole people of the United States, and one which demands the recognition of this Government.

Indeed, I doubt if any statesman of the present time would disown the national character of *this* work, but for the fact that the past history of our Government, in its relations to the great subject of internal improvement, has not been altogether satisfactory to our people. It has happened in the history of former administrations and in more profligate times, that the treasure of the nation has been employed to foster private speculations, and to cherish objects mainly if not purely local in their character.

Such abuses of the authority of the Government over works of internal improvement have engendered a suspicion in the public mind that the central Government could not safely be trusted with the control of those measures, and have led some to deny that it has any authority over the subject whatever.

The enlightened and patriotic gentlemen, who will assemble at Chicago, will take good care, by their action, to give no further cause of complaint in that direction.

And, if it be conceded, that it is of national importance to improve the canals between the Lakes and the Atlantic, it surely must be of national importance to improve the canals between the Mississippi and the Lakes. Those canals hold the same relation to the productions west of the Mississippi, that the Erie canal does to the productions west of the Alleghanies.

If it is of national utility to move the latter from the Lakes to the seaboard at the least possible cost, it must be of national utility to move the former from the Mississippi to the Lakes at the least possible cost. The improvement of the Western canals is certainly not *as* important as the improvement of the Eastern ones; but it is *certainly* important. It may or may not be expedient for the nation now to do the work, but it is unquestionably important to the nation that the work be done.

In this connection, gentlemen, you will indulge me in a single observation. That is this: Whatever fund shall be expended by the representatives of *all* the States, justice would seem to require, should be so expended as to preserve, so far as practicable, the present commercial relations of *each* of the States. You will not ask the American Congress to re-model the geography of the American continent.

There are two water-channels connecting the Mississippi with the Lakes. Both are commercial highways. Both are susceptible of improvement. The upper one, by way of the Wisconsin and Fox rivers, naturally accommodates the commerce of Iowa and Minnesota. The lower one, by way of the Illinois river and canal, naturally accommodates the commerce of Missouri. The greatest grain-growing districts west of the Mississippi are found in Iowa and Minnesota.

To-day those States are nearer the Atlantic, by the eastern routes, than Missouri is. Their products can go to those markets at as little cost, and sell at as much profit as hers. There are more than four degrees of latitude between the entrance to the upper, and the entrance to the lower channel.

It is said that two millions of dollars, expended upon the upper channel, would enable such boats as could alone take the products of Minnesota and Iowa down the Mississippi to the lower channel, to pass directly to the Lakes over the upper one.

It is true, that the improvement of the lower channel, and the neglect of the upper one, would not place Iowa and Minnesota farther from the Atlantic markets than they now are. But it is equally true, it would place them farther from those markets than Missouri would then be. Thereafter, freights from St. Louis would have an advantage in those markets, equal to the cost of more than three hundred miles of transportation.

So far as competition is concerned, it would be all the same as if Missouri was transplanted, at the national cost, between those more northern States and their markets. Such a transposition of communities would be a legitimate object of State enterprise. It may not be thought so legitimate when undertaken by the nation.

So far as the movement of commerce is concerned, it would be all the same as if the Wisconsin river was made to flow through the Illinois. The diversion of water-courses is actionable at common law against an individual. It will hardly be thought just in a nation.

It is said, indeed, by some over-prudent gentlemen, that the times are not propitious to public enterprises of this magnitude.

In my own opinion, that must depend, altogether, upon the character of the enterprises themselves. If they will not benefit and strengthen the nation, there never was a time when it was proper for the nation to engage in them. If they will strengthen the nation, there never was a time so proper as this. I see nothing in the times which calls upon the Republic to delay in its progress or refrain from effort.

To my understanding, the times only call upon the nation more imperatively than heretofore, to comfort its own soul, to respond to its own conscience, to speak truthfully to itself, and to do honorably by each other.

I am not regardless of the fearful demands made upon the public spirit, and the public resources. But I am not appalled by them. Suddenly confronted with the out-crop of a monstrous conspiracy against their supremacy, of more than thirty years growth, I see the energies of the people constantly lifting as their necessities pile up before them, and I *know* they will triumph.

And, fifty years hence, when our children shall look upon droves of iron steeds, springing westward from the Mississippi, to drink in the Pacific, and fleets of ships, steaming eastward to salute the Atlantic, I trust they may gratefully remember, that our proudest contribution to the necessities of peace was bravely made under the extremest pressure of war.

I am, gentlemen, very respectfully,

Your obedient servant,

T. O. HOWE.

To Hon. James Robb, I. N. Arnold, and others, Committee.

FROM EDWIN F. JOHNSON, Esq.,

Civil Engineer, of Middletown, Connecticut.

To James Robb, Isaac N. Arnold, F. C. Sherman, and others, Committee, &c., Chicago:

Gentlemen:—The Great Builder in giving form to the portion of the earth between the Atlantic and the Pacific, impressed upon its surface some very marked features—prominent among which are the two immense valleys or basins, the drainage of which is conveyed to the Ocean by the Mississippi and the St. Lawrence rivers. Between these, and leading also to the same Ocean, are several very much smaller basins, differing from each other in magnitude.

The St. Lawrence basin is characterized by the great fresh-water lakes or seas which it holds in its bosom, the three largest of which have an elevation of six hundred feet nearly above the level of the sea. The remaining two, (Lakes Erie and Ontario,) have an elevation, the one of five hundred and sixty-five feet, and the other of two hundred and thirty-six feet above the same level.

The outlets of these lakes present the various phenomena of flowing waters on a large scale, from the gentle motion of the navigable current, through all the gradations of rapids, cascades, and falls, to the stupendous cataract of one hundred and sixty feet vertical, which for all time has sent forth its unceasing roar between the Lakes Erie and Ontario. This great chain of waters unites with the ocean in a latitude so far to the north, that the ocean-harbors in connection with it, are obstructed by ice during a large portion of the year.

The other great basin, that of the Mississippi, differs very greatly in character from that of the St. Lawrence. It is marked by its vastly greater extent, by the uniformity of its descent or slope, and by the rivers that traverse its surface, of a navigable character, unparalleled (considering their number and extent) in any other equal portion of the globe. These rivers find their union with the sea, through the common channel of the Mississippi, in the region of the tropics, nearly three thousand miles from where the St. Lawrence mingles its waters with those of the same great Ocean.

For more than one thousand miles, commencing at the sources of the Mississippi, in Minnesota, and thence to those of the Alleghany in New York, the several tributaries of these two basins have their rise upon the same elevated ground, situated near to the Lakes and not greatly exceeding them in elevation. Eastward of the Alleghany the southern tributaries of the St. Lawrence interlock with those of the lesser basins, of which mention has been made.

Of the Great Lakes, one lies wholly and the other partially within the limits of the United States. The St. Lawrence river proper is also situated partially within the same limits, as far north as the latitude of forty-five degrees.

The great basin of the Mississippi, covering a surface of one million of square miles, lies wholly within the limits of the United States, except a very small portion near the sources of the Missouri.

The interlocking of the tributaries of the two great basins as described, presents favorable locations in the States of Wisconsin and Illinois for connecting their navigable waters by artificial channels.

A similar favorable location is also presented for uniting the waters of Lake Ontario, with the Hudson, in the State of New York. A connection at these points has already been made, but not in a manner as perfect as the interest of the country requires, by the Wisconsin and Fox river improvements in Wisconsin, the Illinois and Michigan canal in Illinois, and the Oswego and Erie canals in New York.

Such has been and still is the rapid development and growth of our country, that these improvements are fast becoming inadequate, in a commercial and military view, for the purposes contemplated.

Between the Upper Mississippi, the Rocky mountains, and the northern international boundary, is an area of full four hundred thousand square miles, the larger portion of which is adapted to the culture of wheat and other grains, and to grazing, and this character, (such is the direction of the isothermal lines and milder character of the climate towards the West), extends to the region north, including the valley of the Saskatchawan, in the British Possessions.

All this vast region, now trodden, with but few exceptions, only by the foot of the buffalo hunter and the Indian, is destined ere long to be occupied by an intelligent and industrious population, who will there find homes of comfort and peace.

The products of this region will find their cheapest mode of transit to the Eastern States and the Atlantic, by the Great Lakes. The latter will be reached from the valley of the Saskatchawan by the navigation of that river, the Red river of the North, and the St. Peter's branch of the Missouri, which can easily be connected by two or three short canals, having a small amount of lockage. The products of this region, including the Upper Mississippi as far down as the south line of Iowa, seeking the cheapest conveyance to the eastern markets during the season of navigation, will find their way to Lake Michigan through the Wisconsin and Fox river improvement. Those of the great valley of the Missouri, including the Mississippi as far down as the Ohio or below, will find their cheapest navigable route to the great markets of the

East by the Illinois and Michigan improvement, a trade which will eventually, and at no very distant day, be augmented by contributions from beyond the Rocky Mountains, from the valley of the Columbia, and from the Pacific.

It is obvious that the two communications between the Mississippi and Lake Michigan, will not materially interfere with each other. They will command separately an amount of business sufficient to tax their utmost capabilities. To render them adequate to the demand which will be made upon them, and suited to the passage of river-vessels of the larger class, their channels and locks should be enlarged so that these vessels may reach, without change of bulk, the navigation of the Lakes.

This improvement or enlargement, now already required for commercial purposes, is also demanded as a military measure, to enable Government to transfer, whenever required by the exigences of the country, its floating bulwarks from one series of waters to the other.

That these improvements may be productive of the greatest amount of benefit, and that the augmented traffic of the Lakes may find its cheapest and best channel to the leading mart upon the Atlantic, a navigation of a character somewhat different perhaps from the one contemplated between the Lakes and the Mississippi is desirable from Lake Erie to the Hudson.

This navigation should be adapted to the passage of vessels the most suitable for the cheap transmission of freight to the seaboard. The resolutions of the Legislature of New York, at its last session, provide for an increase in the width of the enlarged Erie and Oswego canals of *ten* feet. The locks to be 200 by 26 feet. These dimensions are less than would probably be adopted, if the question was an open one, that is, if there was no canal already in operation between the same points. It is probable that a larger canal would be so difficult of attainment, that the benefit desired might not be an equivalent for the expense incurred in effecting it.

A canal of the dimensions proposed, while it will greatly promote the commercial interests of the country, will fulfil, if not the most perfectly, yet to a very desirable extent, the conditions required of it in a military view.

Intimately connected with this latter improvement, and essential to it, is the construction of a channel suited to lake-going vessels, of all dimensions, or the adoption of some other equally efficient and adequate means for the transfer of vessels and their cargoes, between Lakes Erie and Ontario, within the limits of the State of New York, a channel which should be as far removed as possible from danger or injury by a neighboring power occupying towards us a hostile attitude.

Many years since I had occasion to make a professional examination and report upon the propriety of improving the communication between the Lakes and the Hudson, by the construction of a canal similar in character to that now contemplated.

It was then shown that the cheapest navigable route between

the Hudson and the country bordering the Great Lakes, embraced the waters of Lake Ontario, from Oswego west, and the connection of that lake with Lake Erie, in the manner now proposed, and subsequent experience and observation have confirmed the correctness of the conclusion.

The State of New York has, within the last twenty-five or thirty years, expended large sums in the enlargement of the long line of canal from Lake Erie to the Hudson, and from the latter line to Lake Ontario; but the great and rapidly growing West is still without the benefit of the cheapest and best navigable communication with the Atlantic.

Those conversant with the character of the products to be conveyed between the East and the West, and the cost of conveyance by different modes of transit, will not need to be informed that the great chain of inland waters, stretching, with the few interruptions named, from the Atlantic to the Rocky mountains, in the particular direction which will yet be pursued and adopted for the great national highway from the Atlantic to the Pacific, must perform a most important part in the transmission of those products—must perform a most important part in facilitating commercial exchanges throughout the entire distance named.

It is only by means of the cheap navigation of the great lakes and rivers named, that the heavy and bulky products of the West, when those products are sent from the more remote points, can be delivered at our ocean-marts, at a cost that will insure a reasonable return to the producer.

The improvements proposed will not interfere materially with, but on the other hand will probably constitute an important auxiliary, to the great system of railway intercommunication, which is spreading its iron arms over every portion of our country. As beneficial and useful as is the latter, it cannot supply the place of natural waters for long distances, where time is not an important element, and where cheapness of transit is a ruling consideration.

It will be seen that the improvements in question are truly national in their character, and as such should receive the support of the patriotic and intelligent of all sections of our common country, and if necessary for their accomplishment, they should also receive aid to a reasonable extent from the General and State Governments.

It is for such purposes, among others, that free governments are instituted, to protect and render most effective the labor and industry of the country; to promote in the best, the most just, and most beneficent manner, the general welfare; and to do, within constitutional limits, whatever is required for the best good of all, when the object to be attained cannot be accomplished by individual or municipal agency.

With much respect, I am, gentlemen, very truly yours,

EDWIN F. JOHNSON.

MIDDLETOWN, CONN., May 22, 1863.

FROM HON. W. P. SHEFFIELD,

Late a Member of the House of Representatives from Rhode Island.

NEWPORT, R. I., May 7, 1863.

GENTLEMEN:—By this day's mail I am in receipt of your favor of the 24th ult., inviting me to be present at your meeting to be held in Chicago on the second proximo, to further the enlargement of the canals between the Valley of the Mississippi and the Atlantic coast.

If, by going to Chicago, I could render any practical aid to the project you have in view, I should cheerfully make the journey, for I believe the proposed improvement to be an undertaking of very great national and commercial importance. Its completion would create a new bond of union between the East and the West—cheapen the cost of transporting food, a benefit which would be shared by the producer at the West and the consumer at the East —would stimulate the agricultural enterprise of the West, and induce the culture of other products which would find a ready market in the East. It is believed that indigo, sumac, madder, and other articles, now brought from foreign countries, might be successfully cultivated in parts of the Valley of the Mississippi, which this channel of trade would put in cheap communication with the markets of the East.

I am a believer in the constitution of the United States; but I have but little sympathy in opinion with those men who find, under the power to regulate commerce with foreign nations, the authority to erect and maintain light-houses, and to send expeditions to the Dead Sea, in search of ancient Sodom, and who cannot find in the power to regulate trade between the States, the authority to construct a road, or widen a canal, over or through which the commerce of a great section of the country will necessarily be carried.

The more I reflect on this improvement, the more I am satisfied of its importance, not only to the West, but to the East.

We, of New England, are for developing the resources of the West, for this adds to our national prosperity; but not only for this—for our kindred, and those who are now our neighbors, are of you; and though we are now in the full tide of our strength, if, in after times, adversity should come upon us, we should look to you of the West for support, relying upon your good offices, as a devoted parent would, in times of trial, look for the aid of a worthy son. You are bound to us by ties stronger than those written on parchment; our blood flows in your veins, and the ashes of your ancesters moulder in our soil.

I am but a private citizen. I could do you no good by attending your meeting. Professional employment presses upon my time. I therefore feel constrained to decline the kind invitation with which you have honored me.

With great respect, I am your ob't serv't,

WM. P. SHEFFIELD.

Messrs. JAMES ROBB, I. N. ARNOLD, and others.

FROM THE HON. GEO. OPDYKE,

Mayor of New York City.

MAYOR'S OFFICE, NEW YORK,
May 27, 1863.

HON. ISAAC N. ARNOLD:

DEAR SIR:—Your favor of the 12th inst., inviting my attendance at the National Convention, to be held at Chicago, on Tuesday next, has been duly received. I thank you for the courtesy of this invitation, and the very kind language in which it is conveyed.

I heartily approve of the movement of this Convention, the object of which has my warmest sympathy.

The pressure of business here will, I regret to say, deprive me of the pleasure of meeting with you. I have, however, requested the Common Council of this city to send a delegation to represent us in the Convention, and I trust this will be done at the meeting of the Boards to-morrow.

I inclose a copy of my communication to the Common Council on the subject.

With great regard, I am truly yours,

GEORGE OPDYKE.

FROM HON. S. D. CARPENTER,

Of Wisconsin.

MADISON, May 16, 1863.

GENTLEMEN:—Your kind invitation to be present at the National Canal Convention, to be held at Chicago, on the first Tuesday of June next, was duly received, and in response to your request, I drop you this reply.

I regret that my health and business is such that I cannot be present at your Convention, for nothing could scarcely afford me more pleasure. I assure you, I feel the most lively interest in the expressed objects of the Convention, and cannot but hope that its deliberations will be characterized by prudence, wisdom, and firmness, which alone can secure success.

The object of uniting the Great Lakes with the Mississippi, for commercial and national purposes, concerns not alone the great West, but nearly the whole Union. Facilities for national defense and protection should be the first object of an enterprising people. The commercial advantages to come, as incident, would become incalculable. These links of intercommunication, these channels that float commerce, and keep alive industry, and stimulate enterprise, form the chain which is most needed to strengthen and fasten together the great end and purposes of fraternal union.

The Erie canal not only accomplished much to people the West and enrich the East, but it accomplishes more for civilization, the spread of union sentiment, and all those interests that make us a homogeneous people, than armies could do, or an age of proselyting perform. It has been aptly observed that a successful telegraph-cable between Great Britain and America would keep the peace between the two nations for a century at least. A channel of commerce, between otherwise antagonistic people, is a link of friendship formed by a community of interests, which nothing, save the most adverse wind of fortune, can shake or move.

That the proposed connection between the Lakes and the Mississippi will be the best guarantee of the future peace, not only between the agricultural West and the manufacturing and commercial East, but as between Great Britain and the National Union, I have not the slightest doubt.

The subject is rich with facts and convincing illustrations, but a time is not now allotted to me, I will be content with dropping the above hints, and again assuring you, gentlemen, that though I cannot probably be present in person, I will be with you in spirit and hope, even unto the end.

Very respectfully, yours, etc.,

STEPHEN D. CARPENTER.

To Messrs. Robb, Arnold, Sherman, and others, Committee of Invitation.

FROM CHARLES LANMAN, Esq.,

Librarian of the House of Representatives, Washington.

Washington, May 19, 1863.

Gentlemen:—I have received your invitation to attend the National Convention to be held in Chicago, on the second of June next. The honor thus conferred upon me, I probably owe to my humble connection with the press; but as I am a native of Michigan, and have ever felt special interest in the welfare of the North-western States, I think I am entitled to wish the proposed Convention all the success it deserves. I do this most heartily, and regret that it will not be my privilege to be present on the occasion.

Nature itself has done quite as much for your splendid region of country as for any other, and it is only right that science and art should exert themselves to the utmost, to assist the necessities of nature. In times past, and especially during the present deplorable rebellion, the people of the North-West have proven their patriotism to be unselfish, and sincere, to an unexampled degree; and anything that can be done by the General Government for their prosperity, ought not to be neglected. And as to the importance to the whole country, of greatly increasing the effectiveness of our works of internal navigation, that is a question that does not need any argument.

My first impulse, on receiving your invitation, was to comment somewhat fully upon the important measures which will occupy the attention of the Convention, but men of mark and ability will do all that in person, before the assembly; and they will not, I am certain, forget the eminent dead, whose names are identified with the history of canal-navigation in the United States, and with the prosperity of the Great West. The spirit of DE WITT CLINTON, for example, will undoubtedly be with the members of the Convention, and they should gather encouragement from the perseverance of that great man. Nor will HENRY CLAY be forgotten, who said many things pertinent to your enterprise, and the following was one of them: "No man, who has paid the least attention to the operations of modern war, can have failed to remark how essential good roads and canals are to the success of those operations." And DANIEL WEBSTER, too, whose repeated efforts in behalf of internal improvements the West can never forget, would he not have been among the foremost to advocate the enlargement of the canals at the present time? In a speech that he delivered in the Senate, on the Louisville canal, he uttered a few thoughts which apply with great force to the enterprise now attracting the attention of the statesmen, commercial men, and stalwart yeomanry of the West:

"It is enough for me," said he, "to know that the object is a good one, an important one, within the scope of our powers, and called for by the fair claims of our commerce. So that it be within the Union, it cannot be too remote for me. This feeling, so natural, as I think, to true patriotism, is the dictate also of enlightened self-interest. Were I to look only to the benefits of my immediate constituents, I should still support this measure. Is not *our* commerce floating on these western rivers? Are not *our* manufactures ascending them all, by day and by night, by the power of steam, which is incessantly impelling a thousand engines, and forcing upwards against their currents, hundreds of thousands of tons of freight? If their cargoes be lost, if they be injured, if their progress be delayed, if the expense of their transportation be increased, who does not see that all interested in them become sufferers? Who does not see that every producer, every manufacturer, every trader, every laborer, has an interest in these improvements?"

Thanking the Committee on Invitations for the honor conferred upon me, I have but to express my belief and hope that the objects of the National Canal Convention will be fully and speedily realized.

Very respectfully, gentlemen, your obedient servant,

CHARLES LANMAN.

To the Hon. Committee on Invitations for the National Canal Convention, Chicago, Ill.

FROM GEORGE W. CURTIS, Esq.,

Of New York.

NORTH SHORE, STATEN ISLAND, N. Y.
May 23, 1863.

MY DEAR SIR:—I have left long unanswered your most cordial and friendly invitation to attend your Convention. It will be impossible for me to come, and equally impossible to withhold the heartiest sympathy for a plan so simple, so national, and so worthy and characteristic of the North-West, as the junction of the Mississippi and Lake Michigan.

Gen. Webster's concise, comprehensive, and lucid report seems to exhaust the question. Every great work, which binds every part of us close to every other, is a work dear to every loyal citizen; and the Union is ever secure, so long as we follow the hints of nature, and use every fresh resource of science. Empires weaken by extension; but steam, and the telegraph, and every rapid, and convenient, and obvious means of communication, annihilate distance.

Very truly yours,
GEO. WILLIAM CURTIS.

To Hon. I. N. Arnold, Chicago.

FROM FREDERICK LAW OLMSTED, Esq.,

Secretary of the U. S. Sanitary Commission, Washington.

Washington, May 22, 1863.

My Dear Sir:—I am well convinced, that the purpose intended to be furthered by the meeting at Chicago, on the second of June, is one of the greatest importance to our interests as a nation. I am unable to judge what would be the best method to bring about the accomplishment of that purpose, but I am sure that whatever promises *the broadest and deepest water, in the shortest time*, between *lake and ocean*, that plan is best and in the end will be the cheapest.

I had, until within a few days, hoped to be able to meet you at the Convention, but now find it will be necessary for me to be in New York on the appointed day.

Sincerely regretting this, and thanking you for the invitation,
I am, dear sir, most respectfully yours,
FRED. LAW OLMSTED.

To Hon. I. N. Arnold, Chicago.

FROM MAJ. GEN. S. R. CURTIS, U. S. A.

Headquarters Department Missouri,
St. Louis, May 9, 1863.

Gentlemen:—I am in receipt of your favor of the 20th ult., inviting me to attend the Convention which will convene on the second proximo, "to co-operate in such measures as shall tend to promote the development, security, and unity of the whole country."

Your object is to promote the great idea of uniting the waters of the Mississippi with those of the Lakes, and the St. Lawrence, and Hudson, by such a channel as shall secure profitable and commodious navigation for boats.

The interests involved are so vast, and the difficulty to be overcome so small, I am not surprised at the zeal manifested to secure an immediate effort to accomplish the success of the work. It is only to enlarge works already constructed. It is no matter of doubtful experiment, or novel intricacy. The enlargement of a canal, and the construction of a dam or two on the Illinois river, with suitable locks, can be readily estimated, and soon accomplished.

When this is done, the great rivers of the West will all receive a new and wonderful augmentation of their commercial consequence. Grain and cattle that now find no market, or must be nearly exhausted in value, by transportations and shipments on railroads, will then find such certain and cheap conveyances to the ocean, through healthy climates, as to nearly double their value at home. This will stimulate the increase, and railroads will not be impaired, as they will still have more than they can do, in carrying through light freight, that seeks a more rapid transit.

But it is not in my power, in this letter, to even present the leading elements of such an enterprise. It is proper to press the matter now, because this work is of great national and *military importance;* and our people are now, more than ever, sensible of the necessity and importance of our great national organizations, and the certainty and reality of war. Every link, that binds us as a nation, will now receive special consideration. The commercial union contemplated by you is, therefore, one of great ideas, incident to our terrible revolution. We see new necessities and new objects of pursuit. All natural and artificial combinations deserve and receive new inspiration. The subject of uniting the Mississippi and the Lakes is now easily comprehended, and I cordially approve, and will earnestly advocate, any reasonable plans to accomplish your success.

As other cares will not enable me to attend your Convention, I avail myself of this means of informing you of my hearty cooperation, and my thanks for the honor of your invitation to be present with you on the occasion.

I have the honor to be, gentlemen,

Your very ob't serv't,

S. R. CURTIS,

To Messrs. James Robb, I. N. Arnold, F. C. Sherman, and others, Com., etc.

FROM HON. O. B. FICKLIN,

Late a Member of the House of Representatives from Illinois.

Charleston, May 29, 1863.

Hon. Isaac N. Arnold, and others:

Dear Sirs:—I have the honor to acknowledge the receipt of your letter of invitation to attend the National Convention to be holden in Chicago, on the 2nd of June. The union of our whole country, from centre to circumference, by ship-canals and railways, is of the first importance in a social and commercial point of view, and I doubt not that the proceedings of the contemplated Convention will materially tend to the promotion of that desirable object.

We cannot, in peace or in war, over-estimate the value of intercommunication by canal and railway, between every portion of our common and glorious country, and I may add that the Great West is not without her share of the blessings and responsibilities connected with this enterprise.

One flag and one destiny should, in my humble judgment, be desired by the people of this entire continent, with equal rights and well defined, to every citizen, high or low.

To make the bosom of our great lakes and canals, together with the channel of the Mississippi, the highway of nations for first-class vessels from the St. Lawrence to the Gulf of Mexico, would be a step in the right direction, of priceless value to us.

That your Convention may be a crowning success, in developing the resources of our whole country, is my earnest desire.

Home engagements forbid my absence on the day of your meeting. Very respectfully,

ORLANDO B. FICKLIN.

FROM HON. ERASTUS FAIRBANKS,

Ex-Governor of Vermont.

St. Johnsbury, Vt., May 26, 1863.

Sir:—Other engagements will prevent my responding to the invitation of your Committee to attend the National Canal Convention on the 2nd of June, but I avail myself of the opportunity to express my appreciation of the pro posed enterprise, and my earnest desire that measures may be adopted for a successful prosecution of the work.

I need not speak of its importance. Whether viewed in relation to its military necessity, or its commercial consequence, it is truly a *national work*, and its importance cannot be over-estimated. It is destined to be an essential, if not an indispensable link, in the chain which is to perpetuate the unity of the States, and facilitate not only their commercial intercourse, but their defense against foreign aggression.

The *time* is auspicious for prosecuting the work. It is well understood that the superabundant products of the Western States, in many parts, is rendered comparatively worthless for want of cheap means of transportation to available markets. The financial condition of the country is favorable for undertaking the enterprise.

The events of the war have demonstrated the importance of our inland waters for naval operations, and the public mind is awake to the necessity, as a national measure, of extending these facilities, and of opening a ship-communication from the Mississippi to the

Northern Lakes and the Atlantic ports. These and other considerations commend it to the earnest attention of the intelligent public, and demand for it the patronage of Congress.

I have the honor to be, very respectfully, your obedient servant,

ERASTUS FAIRBANKS.

JAMES ROBB, ESQ., Chairman of the Committee of Invitations, Chicago.

FROM B. J. LOSSING, ESQ.,

Of New York.

POUGHKEEPSIE, N. Y., May 25, 1863.

GENTLEMEN:—I have the pleasure to acknowledge the honor of an invitation from you to attend the National Convention, at Chicago, on the 2nd of June. I sincerely regret my inability to comply, for every man who shall give his voice and influence in favor of the great measures proposed there to be inaugurated, will, I am persuaded, have a record to be proud of.

An industrious and intelligent population, comprising at least one-third of the 100,000,000 of inhabitants which the Eleventh National Enumeration will doubtless exhibit, will occupy immense Free States within the entire latitudinal boundaries of the Republic westward of the Mississippi; and it seems to me that wise statesmanship and enlightened patriotism must necessarily declare that preparations for the proposed facilities for industrial exchanges between the Mississippi valley and the Atlantic seaboard cannot be too soon commenced.

I believe that we are standing, as a nation, on the threshold of a career more glorious and beneficent than has ever been conceived. This war will eradicate the great national disease, vindicate the claim for republican institutions of strength and stability, and prepare the way for the great exodus of the crowded population of the Eastern hemisphere.

The Pacific railways will bring to the Mississippi valley the treasures of Old Cathay; and these and the agricultural and mineral productions of the Great Basin and the regions beyond will demand an almost indefinite expansion of the arteries, through which life-giving internal commerce now circulates.

There is a moral sublimity in your movement at this time, an exhibition of faith in the strength and perpetuity of the Government and of the territorial integrity of the Republic, which must command the admiration of the world. You propose, in the midst of a fierce and gigantic war, by which the very existence of the nation is menaced, measures of vast importance and magnificent proportion, for the aggrandizement of that nation and the good of mankind.

Because of the considerations here hinted at, and a score of others that might be named, I very cordially and hopefully God-speed the enterprise!

With sentiments of high regard, I am, gentlemen, your friend and obedient servant,

BENSON J. LOSSING.

To Messrs. JAMES ROBB, I. N. ARNOLD, and others, Committee on Invitations, Chicago.

FROM HIS EXCELLENCY, J. A. ANDREW,

Governor of Massachusetts.

EXECUTIVE DEPARTMENT, BOSTON,
May 30, 1863.

HON. I. N. ARNOLD, Chicago, Ill.:

DEAR SIR:—I sincerely and deply regret to be obliged to reply to your invitation to attend the National Convention, at Chicago, that pressing and imperative duties, which I cannot postpone or avoid, will detain me in Massachusetts, and prevent my attendance. I beg leave again to express, however, the sincere interest which I entertain in those great projects of internal improvement, which are to form the subject of the deliberations of this Convention, and my sense of the transcendent importance of the various national, political, and economical questions, which are more or less involved with the completion of these grand projects.

I beg to assure you of the great disappointment that I feel, at being deprived of an excursion which I had confidently promised myself, and which I greatly need, and at the loss of an opportunity of seeing personally many of our Western friends, whom it would have been grateful to meet; as well as of enjoying and profiting by the instruction, which we could not fail to find, where so many thoughtful and earnest men will be assembled, for council and deliberation upon matters of such grave importance. But my constant and unremitting attention is imperatively demanded by official duties, from which I cannot escape if I would, and I must reluctantly abandon the hope of being with you on this occasion.

With the most earnest hope that the labors of the Convention may lead to happy and practical results,

I have the honor to be, very respectfully,

Your friend and servant,

J. A. ANDREW.

EXTRACT FROM A LETTER OF HON. J. P. KENNEDY,

Formerly Secretary of the Navy.

BALTIMORE, May 29, 1863.

I am greatly impressed with the importance of the projected enterprise, which, both in value and in magnitude, may be regarded as one of the most brilliant schemes of national statesmanship ever presented to the country—worthy of imperial fame, and destined to produce the most magnificent results in the development of the power of the nation. * * * *

I regret that I cannot be there, even as a spectator; much more, that I cannot take my share in the duty of commending this project to the public attention.

Very truly, yours,

JOHN P. KENNEDY.

JAMES ROBB, Esq.

FROM REV. DR. TAPPAN,

Late Chancellor of the University of Michigan.

UNIVERSITY OF MICHIGAN,
June 1, 1863.

Messrs. JAMES ROBB, I. N. ARNOLD, and others, Committee of Invitation:

GENTLEMEN:—I thank you for the honor of an invitation to attend the National Convention, which meets at Chicago to-morrow. I regret very much that my employments are such as to prevent my attendance. My regret, however, arises from what I shall myself lose, rather than from the hope that my presence would add anything to the practical wisdom which will be there collected.

A Convention, having in view such stupendous public works, designed for the prosperity belonging to times of peace, still more than for the emergencies of war, assembled during the fearful struggle for national existence, is truly a sublime spectacle. The project of connecting the Atlantic with the Mississippi, by ship-canal, is kindred to that of connecting the Atlantic with the Pacific, by a railroad. Born at this time, they evince the deep consciousness of our national strength, the unswerving faith in our national destiny. The purpose to bind together the remotest parts of our great country, by artificial rivers and roads of iron, rising above the waves of civil discord and the storm of battle, is like Neptune of old, rising from the sea, calmly to assert his divinity, by controlling the angry elements:

"alto
"Prospiciens, summâ placidum caput extulit undâ."

The battle we are fighting is for national unity. The great works we are undertaking are for national unity. Confiding in the justice of our cause, and in the God of justice, we shall win the battle. And, entering upon these works with an energy proportionate to their magnitude, we hope that when the battle is won we shall be found prepared for a noble career, in commerce and the attendant arts, and in whatever belongs to national prosperity and national worth.

I am, gentlemen, your ob't serv't,

HENRY P. TAPPAN.

FROM BRIG. GEN. WEBSTER,

Formerly of the Topographical Engineers U. S. A. now of the Army of the Mississippi.

MEMPHIS, June 3, 1863.

Hon. I. N. ARNOLD, Chicago:

MY DEAR SIR:—I sincerely regret my inability to attend the Canal Convention. My duties here do not permit my absence at present.

I need not say how much I hope for your success in your noble efforts to promote the advancement of the great work which is the only link wanting to complete the most magnificent line of internal navigation and commercial and military transit, in the world.

It has been suggested to me, that in my report on this subject, made last fall, I did not fully indorse the reliability of the estimates of Messrs. Gooding and Preston. I must have failed to make myself understood, for I intended to express, what I certainly felt, the *utmost confidence in the full reliability of those estimates.*

I am very truly, your friend and servant,

J. D. WEBSTER.

FROM HON. MONTGOMERY BLAIR,

Post-Master General of the United States.

WASHINGTON, May 30, 1863.

GENTLEMEN:—I regret that I shall be unable to attend the Convention at Chicago on the 2nd of June, to consider the subject of uniting the Lakes and the Mississippi by a ship-canal. I feel assured, however, that the interest awakened in that great measure of national defense, by the present condition of public affairs, secures its full elucidation by the statesmen who have called this meeting, and by others whom they have invited to attend.

I can add nothing to the very able views presented by Mr. Arnold in Congress last winter, and by the Military Committee of the House of Representatives, at the preceding session. The report of that committee shows that an interior channel of navigation is already opened "from New London to Beaufort, N. C., directly communicating with several of our largest States and cities;" that to make this channel available and safe for ship-navigation in war, as well as in peace, requires only "an enlargement of three short and inexpensive canals of the aggregate length of but seventy-eight and a half miles." "That an interior channel of similar importance can be had by means of the Iberville river, and Lakes Mannpa, and Ponchartrain, and Borgne, from the Mississippi to Pensacola;" "that this would unite all the cities of the West, with all the cities of the Gulf, by an interior and protected channel. The cost of this would be even less than the other, and both might ultimately be extended so as to become one. Thus, an interior line of water-communication can be established from New

Orleans to New York and Boston." Individuals having opened this interior line along the West, and the State of New York having taken up the plan of enlarging her canal, the Government of this Union has but to lend its credit for the short line between Chicago and the waters of the Mississippi to complete the circuit of the East and West by inland navigation. This will not only secure our vast inter-state commerce during war, but give protection to all our great harbors and cities by enabling us to concentrate our iron-clad fleet through inland waters upon any one that may be threatened, East, West, North or South.

Thanking you for the consideration shown me by your invitation, I am, very respectfully, your obedient servant,

M. BLAIR.

Messrs. ROBB and others, Committee of Invitation.

FROM JOHN A. POOR,

A Citizen of the State of Maine.

No. 27 UNION PLACE, NEW YORK,
May 30, 1863.

GENTLEMEN:—I am honored by your invitation to attend a Convention at Chicago, on the second of June next, of those in favor of the enlargement of the canals between the Mississippi and the Atlantic. And I am further requested, in case I cannot attend said Convention, to communicate my views in writing, upon the matters embraced in the call.

Until to-day, I had expected to have been able to attend as one of the delegates of the Board of Trade of Portland, Maine, some of whom are on their way, and whose intelligent interest in the success of your efforts will faithfully represent the prevailing opinions of our people.

Your call seems to limit the object of the Convention to the single purpose of an enlargement of the existing canals between the valley of the Mississippi and the Atlantic ocean—works of obvious value, if not all of them of immediate necessity—yet, it may fairly open the entire question of the internal commerce of the country, and the means of transit between the grain-producing regions of the interior of the continent—the great North-West—and their place of market.

Questions of this character are of interest to all, and must, for years, if not for generations to come, become the most engrossing topics of public concern; from the physical configuration of the North American continent, the limited capacity of its natural channels of trade, and the political difficulties in the way of all efforts at the opening of adequate avenues, by artificial means, to meet the wants of a rapidly increasing business.

Great as is now the internal trade of the country, it is a little only of what it will, in a few years, attain to. The production of food is not, at this time, equal to one-tenth of the capacity of the North-western States, without resort to the artificial stimulants that are common in the British Isles. Besides this, one-half of all the grain raised in the United States is produced at points so remote from market, that its value would be consumed in the mere cost of transportation by the ordinary channels. With the aid of all existing canals and railroads, a bushel of wheat in the North-West is only worth one-half its value in Liverpool, so enormous is the cost of present transportation. The question is, how shall this difficulty be overcome? And it is this question alone, that will engage the time and thoughts of the members of this Convention.

It has seemed to me that the great difficulty lies in the way of outlets *from* Chicago, Milwaukee, and other lake-ports, rather than in the lack of means to bring produce to the lake-shores. Cheaply-built and economically worked lines of sailers, with other means of transit, bring into these great granaries—the lake-ports—more produce than the outlets can economically take away.

What is wanted, are cheap and expeditious means of transit, from the Upper Lakes to the open sea. To secure this most effectually, we must make the St. Lawrence-waters AN OPEN MEDITERRANEAN SEA; so that, from the head of Lake Superior and from Chicago, ships of useful size for navigating the ocean can pass, free of duty, and with dispatch, to the Atlantic ports and Europe, and backward to the same places, fully laden. By this means, you could diminish by one-half the cost of transit for the benefit of the farmers of the North-western States, and indirectly, for the advantage of the entire population of the country.

This is a matter of easy accomplishment, if undertaken in the right spirit and temper. The *English-speaking people* of this continent are, for all commercial purposes, *one people*, holding a territory twice the size of the continent of Europe, capable of sustaining as dense a population as that which now occupies that favored portion of the globe. This territory is held in nearly equal shares by the people of the United States and of the British North American Provinces, lying mainly on opposite sides of this great Mediterranean Sea, formed by the waters of the Lakes and the St. Lawrence.

The laws of commerce disregard political boundaries, and the people of the North-West should have their choice of routes to the open sea. Ships should load at Chicago for any port into which an Atlantic sailer can enter, and by so many routes as can be created, from the St. Lawrence, by the way of Lake Champlain, into the Hudson, by the Ottawa, and by Lake Ontario. The advance in the price of *a single crop of wheat* would pay for making all these routes, from Chicago to the Atlantic, navigable for ocean-going sailing-ships and steamers. Montreal harbor could be made for the trade of New York, what Albany is now; and that too, while the St. Lawrence basin, below the Victoria bridge, should be crowded, like the Thames in our day, from London to the sea, when this continent is as fully peopled as Europe.

From Chicago to the Atlantic, for nearly the whole distance, navigation is as cheap as on the ocean. Short canals and lockage would not detain ships more than the average adverse winds of the Atlantic, so that the transit of goods, to and from Chicago and Liverpool, would be nearly as cheap as to and from New York. At one-tenth of the cost of transportation by railway, such a line of navigation would supply an outlet to the trade of the North-West. To transport a ton of goods, by ordinary highways, costs on an average twenty dollars per one hundred miles. The railroads will perform this service for two dollars, the sailing-vessel for one-tenth of this, or twenty cents per ton. Open a ship-canal by the way of the St. Lawrence to Chicago, and the cost of freight will scarcely, if at all, exceed the cost of transit on the ocean, or the Lakes.

Our great difficulties in this country are political ones; greater than the limited amount of capital in business. Public improvements are mainly dependent on local jurisdictions, provinces, or States, governed rather by sectional aims, than by regard to the higher law of commercial convenience. In the United States, nationality has scarcely been regarded as an object of statesmanship, while State Governments have seized upon the more valuable attributes of sovereignty. The regulation of the currency and of the channels of national commerce have been assumed by the States, which should, beyond all other matters, be under the control of the Government of the Federal Union.

To this undue assumption of rights by the States, incompatible with the National Sovereignty, can be traced the origin of the present atrocious civil war, upon the part of rebellious States. This war, however, has already taught us a mode of supplying a national currency which will never be superseded—a discovery worth more than the cost of the war to the present time. Should it enable the National Government to disregard political boundaries in the construction of public works, looking only at physical and commercial law, this war may yet prove to our nation a great blessing.

The highest statesmanship of our day regards *the English-speaking people* of both hemispheres as one in purpose and in destiny. Such an opportunity for greatness, as that enjoyed by the head of the British ministry, has not before this time been offered to any minister of State. He has only to recognize the obvious duties of consanguinity and good fellowship, to make the union of all who speak the English tongue complete in everything that tends to the advancement of civilization, as they are one in purpose and desire. In this spirit let us act. Let political boundaries form no restraint upon commercial enterprise; and the continent, which it is our good fortune to inhabit, shall display exhibitions of material greatness worthy of a superior race, descendants of the heroic men who wrested this new world from the grasp of their less enterprising rivals, and planted over the broad belt of the temperate zone, from the Atlantic to the Pacific seas, institutions and laws, favorable to commercial freedom and constitutional liberty.

If, however, the time has not arrived when we can treat the English-speaking people of the continent as properly subject to our commercial law—a result not very far distant from our day—when an ocean-tariff shall extend with uniform permission, for the collection of duties from Quebec to the Rio Grande, and upon the Pacific coast, with unrestricted power of internal trade; or, in other words, if the British North American Provinces are not ready to adopt with us *an American Zoll-Verin*, we must make use of our own independent advantages. We can, more cheaply than the Canadians have built theirs, construct a ship-canal around Niagara Falls, and from Oswego to the Hudson, that shall, for years to come, take away from the Lakes the surplus produce of the interior. We should further, with the same broad view, deepen the channel of the St. Clair, and extend this water-line, with a capacity equal to the passage of an ocean steamer, from Chicago to the navigable waters of the Mississippi, so that produce can pass by either route to the sea.

The people of the great Republic of the North American continent have been unexpectedly called upon to deal with great enterprises, vast and undefinable in their extent; and while expending, without discontent or embarrassment, large sums in suppressing insurrection, and guarding against foreign invasion, they have found time to contemplate, as necessary practical measures, a railway from the Missouri to the Pacific, and a line of ocean-steamers from San Francisco to the shores of the densely populated continent of Asia. A further knowledge of the capacities of our country and of the capabilities of its people will ensure for them all full and complete success.

With the highest regards, your obedient servant,

JOHN A. POOR.

To Messrs. JAMES ROBB, I. N. ARNOLD, HENRY FARNAM, and others, Committee on Invitations.

FROM THE CORPORATE AUTHORITIES OF THE CITY OF KEOKUK, IOWA.

KEOKUK, IOWA, May 22, 1863.

Messrs. JAMES ROBB, I. N. ARNOLD, and others, Committee of Invitation, Chicago, Ill.:

GENTLEMEN:—Receiving your circular of April 29, and feeling a great interest in the action of the Convention to assemble in your city on the second proximo, our City Council, representing the interests of our community, and vicinity, passed the following:

"Whereas, There has been a call made for a National Convention, to meet in Chicago, on Tuesday, June 2nd, 1863, to consider the enlargement of the canals, between the Valley of the Mississippi and the Atlantic;

Resolved, That the City Council of the city of Keokuk, appoint a Committee of Communication, consisting of two members, the Mayor to be chairman, to communicate with said Convention, expressing our views of the *importance* and *necessity of said enlargement.*"

The object of the Convention, as we understand it, is the enlargement of facilities for transportation from the North-West to tide-water, and the construction of a ship-canal, at Federal expense, from Lake Michigan to the Mississippi river; both of which objects we regard as important and necessary. The first meets our earnest approval, and we believe, that, by enlarging the locks upon the Erie, Oswego, and Welland canals, the present commercial necessities of that portion of the West, from your city eastward, will be attained, as to eastward transportation; and we cannot doubt that New York and Canada will grant this demand. As we lie west of that improvement, we necessarily feel a more vital interest in the extension of such transportation facilities from your city to the Mississippi river.

As a military measure, we cannot commend it too strongly to the General Government. Should our present struggle, for the sustaining of our Government against domestic foes, and the overthrow of this unholy rebellion, lead us into difficulty on questions of comity and international law, with foreign powers, the importance, utility, and necessity of free water-communication from the *Atlantic to the Mississippi river would be felt by all.* In view of this, which, to say the least, at times looks threatening, shall we, supinely, as a people, await such a result before realizing such a necessity? Against such a course we strongly protest.

As a commercial measure, it has our hearty commendation. We regard it as a more important improvement, for the benefit of our commerce, than any ever before projected in this country, since the formation of our government—one that is of vital interest to the growth and prosperity of the North-West. As we of the North-West are purely an agricultural people, (outside of our large cities, where manufacturing is done only to a very limited extent), we appreciate fully the necessity of water-communication with the sea-board. And while our railroads, as some may urge, afford us limited facilities, we are sorry to have it to say, it is at such rates of transportation as almost amounts to restriction. We heartily commend all that is contemplated in the second proposition, and cannot urge too strongly upon your Convention, the importance and necessity of a ship-canal from Chicago to the Mississippi river.

Trusting you may have a large representation in your Convention, and it may not prove in vain that you have thus assembled, we are

Yours, very respectfully,

GEO. B. SMYTH, MAYOR,
C. F. DAVIS,
Committee.

FROM HON. E. JACKSON,

A Citizen of the State of Connecticut.

MIDDLETOWN, CONN., May 19, 1863.

GENTLEMEN:—I have the honor to acknowledge the receipt of your invitation to attend the Convention to be held at Chicago, on the second of June next; and deeply regret that the infirm state of my health deprives me of so great a pleasure.

In obedience solely to your request for my opinion, I cheerfully add my humble voice to the unanimous approval, which, I have no doubt, awaits this great and magnificent project. It truly represents the spirit of the age, which encounters, undiscouraged, the greatest natural obstacles, and subdues the elements themselves to the service of man. Its immense results, social, commercial, and political, no human foresight can reach; and, if ever completed, it will be worthy of as solemn annual commemoration, as that by which the Venitian Republic proclaimed its dominion over the sea.

Very respectfully, your ob't serv't,

E. JACKSON.

Messrs. JAMES ROBB, and others, Committee, Chicago.

FROM HON. J. W. LONGYEAR,

A Member elect of the House of Representatives from the State of Michigan.

LANSING, MICH., May 28, 1863.

Messrs. JAMES ROBB, I. N. ARNOLD, and others, Committee, etc.:

Your circular letter of 29th April, ult., came in my absence from home, and is now before me.

I shall do myself the pleasure to accept your kind invitation, and hope that much good may arise out of the deliberations of the Convention, and that such measures may be devised, as will really "tend to promote the development, security, and unity, of the whole country."

Respectfully, etc.,

JOHN W. LONGYEAR.

FROM J. S. C. KNOWLTON, ESQ.,

A Citizen of the State of Massachusetts.

PALLADIUM OFFICE, WORCESTER, MASS.,
May, 1863.

GENTLEMEN:—It would give me the greatest pleasure, if it were in my power, to be present, as a listener and observer, in the proposed Internal Improvement Convention at Chicago. Our first great work is to put down the rebellion, and vindicate the right of the Government to exist as a Government. When that shall have been done, there will then remain the paramount duty of so arranging and perfecting all the internal and external relations of the country, that it shall be placed forever beyond the contingency of another rebellion—another war of sections upon each other—another subversion of our great industrial aims and purposes—another sacrifice in civil war, of the gains and accumulations of our industry.

But all of us cannot go into active service in support of the war; and, therefore, there is no reason why all such, as are of necessity out of the field of strife, should not give attention to such plans, projects, and measures contemplated, as, if perfected, will bind the Union more closely together; and while preserving all the parts in their integrity, will give an impulse to larger and still larger development. If this country is ever to be destroyed, and to be plunged into the dark pathway of the vanished nations, it will not be done by outward assailments, but by a destructive warfare of sections of clans, of clubbists in interests as well as in passions, wasting each other in unnecessary, because unnatural, struggles. One of the most potent means that can be employed to avert such disasters, in future, is your proposition *to enlarge and multiply the facilities of communication between the great interior and the ocean coasts.* Whoever looks over the map of the United States will see the evidences of such natural adjustment of the several parts of the Union to each other, that he will be ready to pronounce, in the words of DANIEL WEBSTER, that they were designed for "*one country, one constitution, one destiny!*"

The rugged, and comparatively sterile coasts of the country—especially of the Northern States—need the facilities for a speedy and intimate communication with the rich, fertile, and productive lands of the interior. The great necessity of the West is a market. Both sections are essential to each other. We of the East need the grains which you of the West produce; and you of the West need all the markets you can find for your abundant products; for with you, while there is no limit to your supplies of food, you will see to it also, that there is no restriction upon the demand for those supplies. Western men, who talk of "*leaving New England out in the cold,*" may be asked to state their proposition in another form—that of planting so many acres of corn the less, as New England consumes, and of reducing the amount of their produce, in just the extent of the demands made for it, in her market.

We hear much said, and we talk much ourselves, of the great natural outlet of the magnificent Valley of the West, the Mississippi river, and its 80 to 100 tributaries and bayous—all employed in promoting a most healthful communication between the interior points and the exterior lines of this great and growing empire, whose destiny, it is scarcely too much to hope or believe, is the development and rounding-out, in their full proportions, of the industry of freedom, and the intellectual and moral elevation and improvement of the race as individuals and in communities; so that it shall be an empire of MEN, and not alone of material interests.

Monopolies are adverse to the spirit of our institutions. None of us want to be limited to one dull routine, either of business or of enjoyment. Freedom of choice, the cheapest production, and the readiest sale, are the rules of industrial success; and equally true is the declaration, that "*two markets are better than one.*" We of the East have our manufacturing and commercial centres; and you of the West have your agricultural and mercantile centres. The relations of these centres to each other are those of mutuality, and their action, one upon another, is that of reciprocity. It is, and must be a great question in political economy, how to bring these centres, as representatives of great communities around them, into the easiest and quickest action, without any jarring collisions of interest or of passion. The city of Chicago, sitting in queenly majesty, by the side of an internal sea, out of whose placid waters the sun seems to shoot its morning beams, is one of the most important of these great agricultural and mercantile centres. Situated, as it is, at the most southern point of the great system of northern lakes and rivers, it must be, for one-half of the year, the common highway for the trade and travel between the populous East and the great region that lies beyond Lake Michigan, and around Superior; and will continue to be, when the Atlantic and Pacific shall be bound together with bands of iron, for commerce between Western Europe and the great Eastern Empires of China and Japan.

If the Mississippi river is so far the cheapest and best communication with the Ocean, as to be regarded as a monopoly, why is it that we see upon every modern map of the United States, long lines of railway, stretching towards various points of the Upper Valley of the Mississippi, from the ports of New Orleans, and Mobile, Charleston, and Savannah? What is the purpose of these railways, and what are they reaching after? There can be but one answer to the question. The South wants to drain the great Mississippi Basin of its vast wealth of agricultural and mineral products, and it feels the necessity of possessing itself of them in a shorter and more expeditious manner, than by the slow and circuitous route of the Mississippi river, giant as it is among rivers, for the heavy burdens it bears. The East also has its railways grappling the Mississippi Valley; but it feels that something more is wanted, as of "*great national, commercial, and military importance to the country.*" It has the advantage of thousands of miles in the route from the Mississippi to the Ocean and to Europe. That is not enough; it wants a *cheap* as well as an expeditious route. It wants a WATER track as well as an IRON track, and for the simple reason, that while the cost of the *iron* track, originally and continuously, is, of necessity, an immense expenditure, the *water* track, when once constructed, *never wears out.* The difference in the expense of operating the two routes is manifestly great.

While railways between the East and the West will always be preferred for travel and for light and quick freight, a water-communication intermediate between the Mississippi and the St. Lawrence direct, easy, and of sufficient capacity for large operations, both in peace and in war, seems to have been left, by Divine Providence, for the employment of the inventive genius, the constructive skill, the industrial power, of a great people, whose progress is to be attained by a combined force of physical, intellectual, and moral activities.

I have no means of judging of the feasibility of the plans of improvement which are in contemplation, nor their cost, nor of the extent to which such improvement would be generally, or even locally useful. I only know the general fact, that we want, if we can have it, *an ample* WATER-COMMUNICATION, *over the shortest route possible, between the Mississippi and the sea-ports of the East.*

A slight survey of the great physical features of the United States is an assurance that "*the development, prosperity, and unity of our whole country,*" should be the ambition alike of East and West, of North and South; since every consideration of national progress, strength, and unity urges the whole people as with an irresistible logic to find their highest prosperity and happiness in a common brotherhood of sentiments, of rights, of duties, and of obligations.

Very respectfully,

J. S. C. KNOWLTON.

JAMES ROBB, Esq., and others, Committee of Citizens of Chicago.

FROM J. W. BEEKMAN,

A Citizen of the State of New York.

NEW YORK, May 29, 1863.

MY DEAR SIR:—Very greatly to my regret, I find myself compelled, unexpectedly, to abandon my visit to Chicago to-day. I had written to you, some days ago, to say that I hoped to be at the Sherman House on Monday night.

The object of your proposed Canal Convention is of great moment to New York. Before the construction of our Erie canal, the cost of transporting a ton of freight between *Buffalo* and *Albany* was one hundred dollars, and the time of transit was twenty days. In 1858, a ton of merchandise was moved from Buffalo to Albany for two dollars and eighty cents. In 1862 the cost was materially less. The quantity of merchandise moving upon the canal is already reckoned by millions of tons. We ought not to be content until our connection with the great wheat-producing West is as free as it is now between Albany and New York. This city could well afford to pay the interest on the cost of still further enlarging the canal, to a dimension that would admit a Chicago propeller to pass direct between the two cities.

Every public work, which brings the market home to the farm, is an universal blessing. We shall not be always at war. Armed gun-boats and iron-clads may traverse water-ways which peaceful commerce has developed, but *peaceful commerce* must be our aim. Whatever legislation shall bring to the wheat-grower in the West, a larger share of the price his wheat commands in Liverpool, is *wise legislation.* The forwarder thrives better upon a great deal of traffic, at a low price, during a long time, than upon the short spasms of energetic business, at very high rates, which the costliness of transportation now creates.

Rest assured, our interests are identical. If the freight of a bushel of wheat, between your city and ours, could be reduced to twenty cents, leaving eighty cents for the farmer, and some twenty for your merchants, we should gain as much at the East as you could.

In every view of the object you aim at, you ought to succeed. Nothing is more patriotic than to promote intercourse between eastern and western states and cities. Nothing tends more powerfully to make us one—one in hatred of the vile conspiracy of ambitious men, which is striving to ruin our country, and to blast the hopes of freedom the world over, and one in heart and power to put down traitors, than *internal* improvements.

Sincerely and respectfully yours,

JAMES W. BEEKMAN.

James Robb, Esq., Chairman, etc.

FROM JEDEDIAH JEWETT, Esq.,

Collector of the port of Portland, Maine.

Custom-House, Portland,
Collector's Office, May 9th, 1863.

Gentlemen:—Your invitation to attend the National Convention, which is to be held in Chicago, on the second day of June next, has been received by me. While acknowledging with thanks its receipt, I am happy also to state, that the Board of Trade of Portland has already voted to send delegates to that Convention, and that a number, not less than ten, will be present from this city.

We hope to see several of our sister cities in Maine also represented; for, although our State, occupying as it does a frontier position, would be less benefited by lines of intercommunication, than many of her more central sisters, yet, we feel that the State, the corporation, or the individual, which, in this hour of attempted disunion opposes from selfish ends the construction of new lines of traffic and transit, that shall bind together more intimately the various States of our beloved Union, will be equally execrated by posterity, with the memory of that distinguished son of the South who, blessed with the tripple gift of genius, statesmanship and eloquence, illiberally enfolded them all within the boundaries of his own State—South Carolina—and taught his followers to nourish and inculcate the selfish dogma of "State Rights," which has now culminated in secession, rebellion, and war; and that, too, against a Government which has been even more than maternal in her exactions and indulgencies.

Having been chosen one of the Delegates from Portland, I expect to have the pleasure, with my associates, of being present at the Convention, on the second day of June next.

I am, with much respect, your ob't servant,

JEDEDIAH JEWETT.

To Messrs. JAMES ROBB, I. N. ARNOLD, and others, Committee of Invitation for the National Convention, Chicago, Illinois.

FROM HON. JOHN WELLS,

A Citizen of the State of Massachusetts.

SPRINGFIELD, MASS., May 25, 1863.

Messrs. JAMES ROBB, WM. BROSS, and others:

I have just received your very kind letter of the 19th inst. June is a very busy month with me, and I shall not be able to leave home long enough to attend the Convention at Chicago. I am sorry, for I should enjoy another visit there very highly, and I am very deeply interested in the advancement of the internal improvements which the Convention is called to promote.

With the experiences of the last two years, and the threatening aspect of our foreign relations, as warnings to us, it is strange that there should be any difference of opinion or purpose, as to the true policy of our Government in regard to internal means of communication and transportation. But, "will it bring grist to my mill?" is apt to be the inquiry, which settles the course of too many of our people, especially of politicians.

JOHN WELLS.

FROM B. B. FRENCH, ESQ.,

Commissioner of Public Buildings, Washington.

OFFICE OF THE COMMISSIONER OF PUBLIC BUILDINGS,
CAPITOL OF THE UNITED STATES,
WASHINGTON CITY, May 27, 1863.

To the COMMITTEE ON INVITATIONS, Chicago, Ill.:

GENTLEMEN:—I had the honor to receive, some time since, your polite invitation to be present at the great Convention which is to assemble at Chicago on Tuesday next. It would give me great satisfaction to comply with your request, but my official duties here forbid it; and they have been of such a nature recently, as to prevent me from devoting a little time, as I intended, to a reply to your request for my "views in writing."

I have only time now to say, that while the bill authorizing the construction of a ship-canal upon the site of the present Illinois and Michigan canal was pending in the House of Representatives I took considerable interest in the able, and somewhat sharp discussions upon the subject. I read the very able and interesting report, made from the Committee on Military Affairs, by Hon. F. P. Blair, Jr., in February, 1862, and I was convinced, beyond a doubt, of the deep policy as well as great necessity of carrying out, as soon as practicable, the vast improvement advocated in that report.

It would be a work of supererogation, if I had the time, for me to go into the history of the initiation, progress, and, as I viewed it, unfortunate conclusion of that magnificent effort to encompass this Union with what an enthusiastic and eminent friend of mine of enlarged views, denominates "*a hoop of iron that will hold it together forever.*" The failure, I hope and believe was only temporary; and when a renewed effort, under the auspices of the great assemblage of patriots and statesmen at your city, on Tuesday next, shall be made, no true patriot can doubt that it will be successful. That it may be so, any proper effort that I can make shall not be wanting.

This brief letter has been written amid many interruptions, and is not what it ought to be; but, with a heart and soul in the project, I trust the Committee will excuse further *words*, and look to future *acts* on my part, which the old adage quaintly says "are stronger than words!"

With earnest wishes for the ultimate success of the vast and eminently important project which you are to consider,

I am, very respectfully, your obedient servant,

B. B. FRENCH.

Messrs. J. W. Smith, and others, of the Committee.

FROM HON. ROBERT J. WALKER,

Formerly Secretary of the Treasury.

Cork, Ireland, April 18, 1863.

Hon. Isaac N. Arnold, M. C., Chicago, Ill.:

Dear Sir:—Here I am in this beautiful city, in glorious old Ireland, so many of whose gallant sons have fallen in our defense, and thousands more of whom now fight the battles of our country. When I think of Shields, and Meagher, and Corcoran, and their brave associates, shedding their blood that the Union may live; when I feel myself surrounded here by friends of my country, and realize how fervently all Ireland desires our success, my heart swells with gratitude for this noble race, and my prayers are, that Providence would crown them with every blessing.

When you received my promise to attend as a delegate the Chicago Canal Convention, little was it then supposed by me, that duty would call me before that time to Europe. So much of my promise, however, as embraced the discussion of the question, will now be redeemed. The project of an enlarged *thorough-cut* canal, uniting Chicago and the Lakes, with the Illinois river and Mississippi, has long attracted my attention. As a Senator of the United States, for many years, from a South-western State, then devoted to the Union, and elected to the Senate on that question, I have often passed near or over the contemplated route, always concluding that this great work should be accomplished without delay. Every material interest of our whole country demands the construction of this canal, and the perpetuity of the Union is closely identified with its completion. It is for the nation's benefit, and should be the nation's work. It will give new outlets to the Mississippi, through the Lakes, to the Ocean, and neutralize that too exclusive attraction of Western commerce to the Gulf, which has so often menaced the integrity of the Union. We must make the access from the Mississippi, through the Lakes to the Ocean, as cheap, and easy, and eventually as free from tax or toll, as to the Gulf, and the flag of disunion will never float again over an acre of the soil, or a drop of all the waters of the mighty West.

It is clear that, centuries ago, the Lakes and the Mississippi were united, through the Illinois and Wisconsin rivers, and we must remove the obstructions, now divorcing their waters, and restore their union, by thorough-cut canals. In a few years, the saving of transportation, in a single year, would more than pay the cost of the work. The increase of population, wealth, products, imports, exports, and revenue, which would follow the completion of this work, can scarcely now be estimated, and it should be accomplished, if for no other reason, as a most profitable investment of capital for the benefit of the nation.

But great as is the importance of these enlarged canals, uniting the Illinois and Wisconsin rivers with the Lakes, other great works, connecting with the East, are indispensable, as the enlarged locks of the Erie, Champlain, Black river, Syracuse, and Oswego, Cayuga, Seneca, Chemung, and Elmira to the Pennsylvania State line, Rochester, and Alleghany river. Nearly all of these are seventy feet wide and seven feet deep, and require only an enlargement of the locks, whilst a few require to be widened and deepened. The Chemung canal connects the Susquehanna with the Erie canal, at Montezuma, and the Chenango is nearly completed to the north branch of the Susquehanna at the Pennsylvania State line, whence, the Susquehanna canal passes through Wilkesbarre, Northumberland, Middleton, and Wrightsville, to Havre de Grace, in Maryland, on tide-water, at the head of the Chesapeake Bay. The great canal, from the southern boundary of New York, down the Susquehanna to tide-water, is now five feet deep, and from forty to fifty feet wide, and can be readily enlarged to the dimensions of the Erie canal. With these works thus enlarged, the connection of the Lakes would not only be complete with the Hudson, and by the Delaware and Raritan canal with the Delaware, and by the Delaware and Chesapeake canal with the Chesapeake Bay, but also by the direct route, down the Susquehanna, to Baltimore, Norfolk, and Albemarle sound. Is not this truly national, and is it not equally beneficial, to the East, and the West, to open all these routes for large steamers? The system, however, would not be complete, without uniting Champlain with the St. Lawrence, Ontario with Erie, and Huron and Michigan with Superior.

The enlarged works should also be provided through Wisconsin, Indiana, Ohio, and Western Pennsylvania, to the Lakes, to the extent that these canals can be made of the dimensions of the Erie, and supplied with water. Nor should we forget the widening of the canal at Louisville, the removal of obstructions in the St. Clair flats, and Upper Mississippi, and the deepening of the mouth of this great river. The construction of these works would be costly, but as a mere investment of capital, for the increase of our wealth and revenue, they would pay the nation tenfold.

As the main object of these works is cheap transportation, the tolls should be diminished, as the works were completed, to the full extent, that freight could be carried more cheaply in large boats, and provision should be made for an adequate sinking-fund, so as gradually to liquidate the whole cost, and then collect no more tolls than would pay to keep the works in repair. Such is the true interest of the States and the nation. If New York could collect a toll for navigating the Hudson, it would be against her interest, for the diminution and diversion of business, and tax on labor and products, would far exceed the net proceeds of any such toll. The same principle will apply to these canals. As some of them, unfortunately, are owned by private companies, adequate provision should be made to prevent these aids from being perverted to purposes of individual speculation. The Erie and Ontario canal, at the falls of Niagara, and the Superior, Huron, and Michigan canal (less than a mile long,) at the Falls of St. Mary, should be made ship-canals, much larger than those of Canada.

The cost of all these works may exceed $100,000,000, but the admirable financial system of Mr. Secretary Chase, would soon supply abundant means for their construction. Already the price of gold has fallen largely, our legal tenders are being funded, by millions, in the Secretary's favorite 5-20 sixes, and we shall soon have, under his system, a sound, uniform, national currency, binding every State and citizen to the Union, and fraught ultimately with advantages to the nation, equal to the whole expense of the war.

In passing down the Susquehanna canal, at Middletown, commences the canal, which, by way of Reading and the Schuylkill, connects Philadelphia with the Susquehanna and the Lakes. Most of this work is already six feet deep, but the whole route, if practicable, should be enlarged to the dimensions of the Erie canal.

I have met in the British Museum some documents showing the original project (absurdly abandoned) for a large canal from the Schuylkill to the Susquehanna. A slight change will restore this work, and give to Philadelphia a complete seven-foot canal, via the Schuylkill and Susquehanna, to the Lakes, as short as from New York, and through a richer country, both mineral and agricultural. It appears that Washington and Franklin both favored this route.

1. Gun-boats, and large commercial steamers could then pass, without interruption, through all the lakes, to the St. Lawrence, the Hudson, the Delaware, Susquehanna, Chesapeake Bay, Albemarle Sound, the Ohio, and Mississippi.

2. In case of war, foreign or domestic, the saving to the Government in prices of articles they must buy, and in transportation of men, munitions of war, supplies, and coal, would be enormous. It is believed that the excess of cost in prices and transportation during this rebellion, occasioned by the want of these works, WOULD MORE THAN PAY FOR THEIR CONSTRUCTION. Nor is this the only loss, but victories no doubt have often been turned into defeats, for the want of proper facilities for the movement of gun-boats, of supplies, and munitions, and the rapid concentration of troops and reinforcements.

3. The ability to obtain supplies, and coal and vessels, from so many points, and especially gun-boats, where the coal, iron, and fixtures are in juxtaposition, would hasten construction, and cheapen prices to the Government.

The enormous naval and military power, gained by such works, would tend greatly to prevent wars, foreign or domestic, or, if they did occur, would enable us to conduct them with more economy and success. It is said such vessels can be built on the Lakes, and so they can, for lake defense, but they would be liable to capture or destruction there, before completed, by the enemy, and iron vessels and iron-clads could not be constructed so cheaply where there is neither coal nor iron, as in regions like the Delaware, Susquehanna, Alleghany, and Ohio, where these great articles abound, and can be used on the spot, with so much economy.

It must be remembered, also, that, if these iron steamers and iron-clads are constructed on the seaboard or the lakes, still the iron and coal for building them, and the coal for running them, could be supplied much more cheaply, if these enlarged canals were finished. Besides, events are now occurring, and may again, in our history, requiring the immediate construction of hundreds of iron vessels, rams, iron-clads and mortar boats, calling for all the works on the seaboard, the Lakes, the Western rivers, and enlarged canals, to furnish, in time, the requisite number. Rapid concentration of forces, naval and military, and prompt movements, are among the great elements of success in war. It will be conceded, that the ability to run gun-boats, iron-clads, rams, and mortar boats, through all our lakes, to and from them to all our great rivers, and to connect from both, through such enlarged canals, with the seaboard, and the Gulf, would vastly increase our naval and military power.

Is it not clear, then, that if such a movement, with such resources and communications, had been made, in sufficient force, the first year of the war, so as to seize, or effectually blockade, all the rebel ports, to occupy, by an upward and downward movement, the whole Mississippi and all its tributaries, isolating and cutting rebeldom in two, and thus preventing supplies from Texas, Louisiana, and Arkansas, that the contest must have been closed long ere this, and thus saved five or six times the cost of these works. As indicating the consequence of our occupation and command of the Mississippi and the Gulf, let us see its effect on the supply of the single indispensable article of *beef* to the rebel army and people.

By the census of 1860, table 36, the number of cattle that year in the loyal States was 7,674,000; in Texas alone, 2,733,267; in Louisiana, 329,855; and in Arkansas, 318,355; in those parts of the rebel States east of the Mississippi, not commanded by our troops and gun-boats, 2,558,000, and in the parts of those States thus commanded by us, 1,087,000. Thus it will be seen, that the cattle in Texas alone (whence the rebels, heretofore, have derived their main supplies,) raised on their boundless prairies, and rich perennial grass, have largely exceeded all the cattle in those parts of the rebel States east of the Mississippi, commanded by them. But that commanded by us, of the Mississippi and its tributaries, and the Gulf, as is now the case, cuts off the above supplies from Texas, Louisiana, Arkansas, west and north Mississippi, north Alabama and west and middle Tennessee. Hence the cries of starvation from the South; hence, mainly, the rise there is in the price of beef, from a few cents to a dollar a pound. Controlled, as the Gulf and Mississippi and its tributaries now are by us, so as to prevent any Western supplies of beef, and the desolation and inundation which has swept over so much of the rest of rebeldom since 1860, their army and people cannot be supplied with beef throughout this year. Nor would running the blockade help them in this respect, for Europe has no supplies of beef to spare, requiring large amounts from us every year. The revolt, then, is doomed this year by starvation, if not, as we believe, by victories. Indeed, I imagine, if our Secretaries of War and of the Navy were called on for official reports, they could clearly show, that with ample appropriations in July, 1861, and all these works *then* completed, they could have crushed the revolt in the fall of 1861 and winter of 1862. All, then, that has been expended since, of blood and treasure, and all the risks to which the Union may have been exposed, result from the want of these works. Surely these are momentous considerations, appealing, with irresistible force, to the heart and judgment of every true American statesman and patriot.

Great, however, as are the advantages in war to be derived from the construction of these works, it is still more in peace, and as arteries of trade, that the benefit would be greatest. If iron steamers are to control the commerce of the world, the cheap construction and running of such vessels may decide this great question in our favor. Now, whether these steamers are to be built on the seaboard or the interior, the coal, and iron, and timber, with which to make them, and the coal and supplies with which to run them, could be furnished much more cheaply by these enlarged canals. And even if the vessels be of timber, the engines, boilers, anchors, etc., must be of iron, and they must be run with coal, all which would be furnished cheaper at our lakes and seaboard, by these enlarged canals. Nor is it only for the construction of engines and boilers for steamers, or coal to run them, that these works would be important, but the cheapening of the transportation of coal, iron, timber, and supplies, would be greatly beneficial in all industrial pursuits. It is, however, in cheapening the transportation of our immense agricultural products to the East, South, seaboard, and the return cargoes, that these works would confer the greatest benefits. The value of the freight transported on these canals, last year, was over $500,000,000, but, when all should be enlarged, as herein proposed, the value of their freight, in a few years, would exceed SEVERAL BILLIONS OF DOLLARS. They would draw from a vastly extended area, from augmented population and products, and with greater celerity and economy of movements, from the increased distances that freight could be carried, and additional articles. With these improvements, millions of bales of cotton would be carried annually on these enlarged canals. All of Missouri, Iowa, Kentucky, Minnesota, Kansas, and the North-western Territories, up the Missouri and its tributaries, with large portions of Tennessee, Alabama, Arkansas, Louisiana, and even of Texas, on the Red river, would be added to the region from which supplies would be sent, and return cargoes proceed by these works. Our exports abroad would soon reach a BILLION of dollars, of which at least one-third would consist of breadstuffs and provisions. Corn was consumed, last year, in some of the Western States, as fuel, in consequence of high freights. But this could never recur with these enlarged canals. Indeed, the products to be carried on these canals would include the whole valley of the Lakes, the Ohio, Mississippi and Missouri, and many articles thus reaching there, thence be carried, on our great imperial railway, to the Pacific, bringing back return cargoes for the same routes. Breadstuffs and provisions and cotton would be carried more cheaply through these canals to the manufacturing States, and their fabrics return the same way, in vastly augmented amounts, to the West.

Last year, even during a war, breadstuffs and provisions, reaching $109,676,875 in value, were exported abroad, from the loyal States alone; but, with these enlarged canals, the amount could be more than tripled, the augmented exports bringing in increased imports, and vast additional revenue. Can we not realize the certainty of these great results, and have we not the energy and patriotism to insure their accomplishment? Assuredly we have.

Nor is it only our revenue from duties that would be increased to an extent sufficient of itself, in a few years, to pay the principal and interest of the debt incurred in the construction of these works, but our internal revenue, also, would be prodigiously augmented.

The census of 1860 shows our increase of wealth, from 1850 to 1860, to have been 126.45 per cent. (Table 35.) Now, if we would increase our wealth only one-tenth, in the next ten years, by the construction of these works, then (our wealth being now $16,159,616,068) such increase would make our wealth, in 1870, instead of $36,593,450,585, more than sixteen hundred millions *greater*, or more than *ten times* the cost of these works; and, in 1880, instead of $82,865,868,849, over three billions six hundred millions more, or more than twenty times the cost of these works. The same per centage, then, of our present internal tax, on this augmented wealth, estimated at only one per cent., would be $16,000,000 (annually) in 1870, and $36,000,000 (annually) in 1880, and constantly increasing. Add this to the great increase of our revenue from duties, as the result of these works, and the addition would not only soon liquidate their cost, but yield a sum which, in a few years, would pay the principal and interest of our public debt.

With such works, we would certainly soon be the first military, naval, and commercial power of the world. The West, with these reduced freights, would secure immense additional markets for her products, and the East send a much larger amount of manufactures, in return cargoes, to the West.

A new and great impulse would be given to the coal and iron interest. If the Delaware, Susquehanna, and their tributaries, and the Ohio and its tributaries, especially the Youghiogheny, Monongahela, and Alleghany had the benefit of low freight, afforded by these canals, they could supply not only the seaboard at reduced rates, but also, central and western New York, the Canadas, and the whole Lake Region, with coal and iron. Indeed, the increased demand, thus caused for these great articles, would soon bring our make of iron and consumption of coal, up to that of England, and ultimately much larger. Freight is a much greater element in the cost of coal and iron, than of agricultural products, but the increased exchange would be mutually advantageous.

With this system completed, the Mississippi might communicate by large steamers, with all the lakes, and eastward by the enlarged canals, to Chicago or Green Bay, or pass up the Ohio, by the Wabash or from Lawrenceburg or Cincinnati to Toledo, or by Portsmouth or Bridgeport to Cleveland, or by Bridgeport to Erie City, or by Pittsburgh, up the Alleghany to Olean and Rochester, on the Erie canal, or by ship-canal, from Buffalo to Ontario, thence, by the St. Lawrence to Lake Champlain and the Hudson, or by Oswego to Syracuse, or by the Erie canal from Buffalo to the Hudson, or by the Chenango and Chemung route, down the Susquehanna, to Philadelphia, or Baltimore, or down the Chesapeake to

Norfolk, and on through Albemarle Sound south. Or, going from the East, or South, westward by these routes, the steamers could proceed west, and up the Missouri, to the points where they would meet the great railway leading to the Pacific. Indeed, if we do our duty now, the next generation may carry similar canals from the head of Lake Superior to the Mississippi and Missouri, and up the Kansas or Platte to the gold mines of Colorado, or, from the great falls of the Missouri, to the base of the Rocky mountains, with railroad connection thence to the mouth of the Oregon and Puget's sound. There would be connected with this system, the Lakes, and all the Eastern waters, the Ohio and all its tributaries, including the Youghiogheny, Monongahela, Alleghany, Kanawa, Guyandotte, Big Sandy, Muskingum, Scioto, Miami, Wabash, Licking, Kentucky, Green river, Barren, Cumberland, and Tennessee, the Upper Mississippi and its tributaries, especially the Illinois and Wisconsin, the Des-Moines and St. Peter's, the lower Mississippi and all its vast tributaries, the St. Francis, White river, Arkansas, Red river, and Yazoo. These are no dreams of an enthusiast, but advancing realities, if *now*, *now* we will only do our duty in crushing this rebellion, and exorcise the foul fiend of slavery, that called it into being. We may best judge of what we may do in the future, by what we have done in the past. We have constructed 4,650 miles of canals (including slackwater), at a cost of $132,000,000. We have constructed (including city roads) 31,898 miles of railroad, at a cost of $1,203,285,569, making an aggregate, for railroads and canals, of $1,335,285,569. Now, one-tenth of this sum will probably make all the works proposed now to be executed, for they are all only enlargements of existing canals, except the ship-canal around the falls of Niagara, and a similar canal from Lake Champlain to the St. Lawrence, a work of vast importance, but that can only be accomplished with the aid and consent of Canada, and is not now estimated.

These improvements would be truly national, especially as provision would be made for deepening the mouth of the Mississippi. We propose to make or enlarge no mere local works, but only such as connect the Atlantic and the Gulf with the Lakes, Ohio, Mississippi, Missouri, Hudson, Delaware, Susquehanna, Chesapeake bay, and Albemarle sound. These local routes must be constructed or enlarged by local or State expenditures.

The canals in New York, constituting so large a portion of the system, have already (mostly) the requisite width and depth, and only need an enlargement of the locks. The great Delaware and Raritan canal, connecting New York with Philadelphia, has a depth of eight feet, and the Delaware and Chesapeake, uniting them with the Susquehanna and the Lakes, Baltimore, Norfolk, and Albemarle sound, has a depth of ten feet. No doubt the enlightened proprietors of the Delaware and Raritan canal would, on fair terms to themselves and the Government, enlarge that canal (if practicable) to the depth of the Delaware and Chesapeake, which would be of incalculable benefit to the whole country, but especially to New York, Philadelphia, Baltimore, and Norfolk.

The Pennsylvania canals proposed to be enlarged, are the Schuylkill, leading by the Union from Philadelphia, through Reading to Middleton on the Susquehanna, and thence up that river to the Erie and the Lakes. The Schuylkill canal, 70 miles to Reading, has a depth of six feet, and from Reading to Middleton, four feet. The Susquehanna canal, from Havre de Grace, Maryland, at the head of tidewater, and the Chesapeake bay to the New York line, and system, has a uniform depth of five feet, and is about 300 miles long. This canal, leading through Maryland and Pennsylvania along the Susquehanna, can readily and cheaply be enlarged to the dimensions of the Erie canal, and will then furnish Norfolk, Baltimore and Philadelphia a direct route to the Lakes by the enlarged system, fully equal to that of New York. Western Pennsylvania and Pittsburgh would have the route, by the enlarged system, up the Allegany and Olean to Rochester on the Erie canal, and thence to the Hudson or the Lakes, and from Bridgeport to Cleveland or Erie City. Ohio would have the benefit of the routes (enlarged) to and from Cleveland to Bridgeport or Portsmouth on the Ohio, and to and from Toledo to the mouth of the Wabash or Miami, or to Cincinnati. These canals are forty feet wide and four feet deep. Indiana would have the benefit of the Wabash and Erie canal to Evansville, on the Ohio, from Toledo, and to and from the same point to the mouth of the Miami at Lawrenceburg and to Cincinnati, and would also largely participate in the benefit of the Chicago and Illinois canal of the whole system. Wisconsin would have the benefit of all these canals, but especially of that connecting the Wisconsin river with Green Bay, and the rest of the lakes with Lake Superior. Illinois would have the benefit of the Wabash and Erie, the Chicago and Illinois, and of the entire system. Indeed, with a thorough-cut canal from the Illinois river to Chicago, fact will outstrip fancy as regards the progress of that great city. And here a strong argument in favor of the whole of these works is presented to every true American, by the fact that the vast and increasing heavy and bulky products of the West demand the enlarged works, and if she cannot have them by the Hudson, the Delaware, and Susquehanna, she will have them by the Canada canals, and the St. Lawrence to its outlet in the Gulf. Minnesota would have the benefit of the improvement of the upper Mississippi, and of the canals uniting the Wisconsin with Green Bay, and Superior with the other lakes. Iowa, Missouri, Kansas, and the whole Western Territories would have the benefit of the improvement of the Mississippi, of the routes by Chicago, Green Bay, the Ohio, and the whole system. The glorious new free State of Western Virginia would have the benefit of all the routes up and down the Ohio and Mississippi to the Lakes, the Hudson, Delaware, Susquehanna, and the Gulf. So would Kentucky, and the enlargement of the Louisville canal would be within her own limits. When we reflect that Kentucky borders for nearly a thousand miles on the Ohio and Mississippi, with her streams, the Big Sandy, Licking, Kentucky, Green river, and Barren (which last four have 766 miles of slackwater navigation), Cumberland, and Tennessee, all tributaries of the Ohio, the benefits to her would be prodigious. The interest of the States of Tennessee, Arkansas, Mississippi, Louisiana, North Alabama, on the Tennessee river, and Texas, on the Red river, would be greatly promoted. They would have improved routes to and from the mouth of the Mississippi, and to and from the Ohio, the Lakes, and the Atlantic. Eastern Virginia and North Carolina would derive great advantages by the enlarged routes, connecting Albemarle sound and the Chesapeake with New York, Philadelphia, the Hudson, the Delaware, the Susquehanna, and the Lakes. Delaware and Maryland could avail themselves most beneficially of all these routes, and Baltimore would derive immense advantages from the enlarged route by the Susquehanna to the Lakes, having then as good a route there as New York, and the difference of distance being only 30 miles. New Jersey, by her route from the Delaware and Raritan to the Hudson, and by her rising cities near or opposite Philadelphia and New York, and by the enlarged system to the Lakes, would find all her interests greatly advanced, and the business on her canals and railroads vastly increased. Michigan, with a larger lake shore than any other State, fronting on Lakes Superior, Michigan, Huron, St. Clair, the connecting straits and rivers, and Lake Erie, would derive immense advantages. By her immediate connection with the whole New York and Eastern system, and by Toledo, Cleveland or Erie City, to the Ohio river, and by the Chicago or Green Bay routes to the Mississippi and the Gulf, her vast agricultural products in the peninsula would find new and augmented markets; while, with the ship-canal to Lake Superior, the magnificent iron and copper mines on that immense inland sea, as well as those in Wisconsin, and the splendid pineries and fisheries of both States, would receive an immense development. Pennsylvania has no large available through route now from the Delaware and Susquehanna to the Lakes, nor from Pittsburgh. The proposed system would give her those routes, as well from the East as from the West. This would give to her coal and iron, her vast agricultural products, her immense manufactures, and all her industrial pursuits, a new impulse, while her two great cities, Philadelphia and Pittsburgh, would be greatly advanced in wealth and population. When we reflect that coal and iron have mainly contributed to make England what she is, and how superior, in this respect, are the natural advantages of Pennsylvania with her bituminous and anthracite coal and iron and fluxes in juxtaposition, with a *continent* surrounding her to furnish a market, with her central location, fronting on the deep tidewater of the Delaware, and upon the Lakes and the Ohio, with its two great confluents at Pittsburgh, the Alleghany and Monongahela, we cannot fully realize the immense advantages which she would derive from these enlarged communications. But what of New York? With all her routes, as well as that of the Erie canal, enlarged as proposed, with her mighty system extending to the Lakes and St. Lawrence, from Lake

Champlain to Superior, south by the Delaware and Susquehanna, west by the Alleghany, Ohio, Missouri, and Mississippi, and her great city with an unrivalled location, what an imperial destiny lies before her, with the Union preserved? Oh! if she would only fully realize these great truths, and spurn from her embrace the wretched traitors who, while falsely professing peace, mean the degradation of the North and the dissolution of this Union, who can assign limits to her wealth and commerce?

Let us now examine the relations of New England to these proposed works. Vermont, upon Lake Champlain, by the enlarged system, connecting her with the Hudson, the St. Lawrence, and the Lakes, would be greatly advanced in wealth and population. But with cheapened transportation to and from Lake Champlain or the Hudson, and not only Vermont, but all New England, in receiving her coal and iron, and her supplies from the West, and in sending them her manufactures, will enjoy great advantages, and the business of her railroads be vastly increased. So, also, New England, on the Sound, and, in fact, the whole seaboard and all its cities. Bridgeport, New Haven, New London, Providence, Fall River, New Bedford, Boston, Portland, Bangor, Belfast, and Eastport, will all transact an immense increased business with New York, Philadelphia, Baltimore, and the West. As the greatest American consumer of Western breadstuffs and provisions, and of our iron and coal, and the principal seat of domestic manufactures, the augmented reciprocal trade of New England with the South and West will be enormous. Her shipping and ship-building interests, her cotton, woollen, worsted, and textile fabrics, her machinery, engines, and agricultural implements, boots and shoes, hats and caps, her cabinet furniture, musical instruments, paper, clothing, fisheries, soap, candles, and chandlery, in which she has excelled since the days of Franklin, and, in fact, all her industrial pursuits, will be greatly benefited. The products of New England in 1860, exclusive of agriculture and the earnings of commerce, were of the value of $494,075,498. But in a few years after the completion of these works, this amount will be doubled. Such is the skilled and educated industry of New England, and such the inventive genius of her people, that there is no limit to her products, except markets and consumers. As New York increases, the swelling tide of the great city will flow over to a vast extent into the adjacent shores of Connecticut and New Jersey; and Hoboken, West Hoboken, Weehawken, Hudson City, Jersey City, and Newark, will meet in one vast metropolis. Philadelphia will also flow over in the same way into Camden and adjacent portions of New Jersey, whose farms already greatly exceed in value those of any other State. The farms in New Jersey in 1860 were of the average value of $60.38 per acre, while those of South Carolina, the great leader of the rebellion, with all her boasted cotton, rice, and tobacco, and her 402,406 slaves, were then of the average value of $8.61 per acre. (Census table 36.) And yet there are those in New Jersey who would drag her into the rebel confederacy, cover her with the dismal pall of slavery, and who cry *Peace! peace!* when there is no peace, except in crushing this wicked rebellion. The States of the Pacific, as the enlarged canals reached the Mississippi and Missouri, and ultimately the base of the Rocky mountains, would be greatly advanced in all their interests. Agricultural products and other bulky and heavy articles that could not bear transportation all the way by the great Pacific railroad, could be carried by such enlarged canals to the Mississippi and Missouri, and ultimately to the base of the Rocky mountains, and thence, by railroad, a comparatively short distance to the Pacific, and westward to China and Japan. In order to make New York and San Francisco great depots of interoceanic commerce for America, Europe, Asia and the world, these enlarged canals, navigated by large steamers, and ultimately toll free, are indispensable.

We have named, then, all the Territories, and all the thirty-five States, except three, as deriving great and special advantages from this system. These three are Georgia, South Carolina, and Florida, with a white population, in 1860, of 843,338. These States, however, would not only participate in the increased prosperity of the whole country, and in augmented markets for their cotton, rice, sugar, tobacco, and timber, and in cheaper supplies of Eastern manufactures, coal, iron, and Western products, but they would derive, also, special advantages. They have a large trade with New Orleans, which they would reach more cheaply by the deepening the mouth of the Mississippi. They could pass up Albemarle sound, by the interior route, to Baltimore, Philadelphia, New York, or the West, and take back return cargoes by the same route. Georgia, also, by her location on the Tennessee river, together with South Carolina, connected with that river at Chattanooga, would derive great benefits from this connection with the enlarged canals and improved navigation of the West, sending their own and receiving Western products more cheaply.

Thus, every State and every Territory in the Union would be advanced in all their interests by these great works, and lands, farms, factories, town and city property, all be improved in value.

But, there is another topic connected with this subject, of vast importance, particularly at this juncture, to which I must now refer. It is our public lands, the homstead bill, and immigration. On reference to an article on this subject, published by me in the November number of THE CONTINENTAL MONTHLY, it will be found that our unsold public lands embraced 1,649,861 square miles, being 1,055,911,288 acres, extending to fifteen States and all the Territories, and exceeding half the area of the whole Union. The area of New York being 47,000 square miles, is less than a thirty-fifth of this public domain. England (proper), 50,922 square miles; France, 203,736; Prussia, 107,921; and Germany, 80,620 square miles. Our public domain, then, is more than eight times as large as France, more than fifteen times as large as Prussia, more than twenty times as large as Germany, more than thirty-two times as large as England, and larger (excluding Russia) than all Europe, containing more than 200,000,000 of people. As England proper contained, in 1861, 18,949,916 inhabitants, if our public domain were as densely settled, its population would exceed 606,000,000, and it would be 260,497,561, if numbering as many to the square mile as Massachusetts. These lands embrace every variety of soil, products and climate, from that of St. Petersburg to the tropics.

After commenting on the provisious of our homestead bill, which gives to every settler, American or European, 160 acres of land for ten dollars (the cost of survey, etc.), I then said:

"The homestead privilege will largely increase immigration. Now, besides the money brought here by the immigrants, the census proves that the average annual value of the labor of Massachusetts, *per capita*, was, in 1860, $220 for each man, woman and child, independent of the gains of commerce—very large, but not given. Assuming that of the immigrants at an average annual value of only $100 each, or less than thirty-three cents a day, it would make, in ten years, at the rate of 100,000 each year, the following aggregate:

1st	year	100,000	=	$10,000,000
2nd	"	200,000	=	20,000,000
3rd	"	300,000	=	30,000,000
4th	"	400,000	=	40,000,000
5th	"	500,000	=	50,000,000
6th	"	600,000	=	60,000,000
7th	"	700,000	=	70,000,000
8th	"	800,000	=	80,000,000
9th	"	900,000	=	90,000,000
10th	"	1,000,000	=	100,000,000
		Total		$550,000,000

"In this table the labor of all immigrants each year is properly added to those arriving the succeeding year, so as to make the aggregate the last year 1,000,000. This would make the value of the labor of this million of immigrants in ten years, $550,000,000, independent of the annual accumulation of capital, and the labor of the children of the immigrants (born here) after the first ten years, which, with their descendants, would go on constantly increasing.

"But by the official returns, (p. 14, Census,) the number of alien immigrants to the United States, from December, 1850, to December, 1860, was 2,598,216, or an annual average of 260,000.

"The effect, then, of this immigration, on the basis of the last table, upon the increase of national wealth, was as follows:

1st	year	260,000	=	$26,000,000
2nd	"	520,000	=	52,000,000
3rd	"	780,000	=	78,000,000
4th	"	1,040,000	=	104,000,000
5th	"	1,300,000	=	130,000,000
6th	"	1,560,000	=	156,000,000
7th	"	1,820,000	=	182,000,000
8th	"	2,080,000	=	208,000,000
9th	"	2,340,000	=	234,000,000
10th	"	2,600,000	=	260,000,000
		Total		$1,430,000,000

"Thus, the value of the labor of the immigrants, from 1850 to 1860, was $1,430,000,000, making no allowance for the accumulation of capital, by annual reinvestment, nor for the natural increase of this population, amounting, by the census, in ten years, to about twenty-four per cent. This addition to our wealth, by the labor of the children, in the first ten years, would be small; but in the second and each succeeding decade, when we count children, and their descendants, it would be large and constantly augmenting. But the census shows that our wealth increases each ten years at the rate of 126.45 per cent. (Census Table 35.) Now, then, take our increase of wealth, in consequence of immigration, as before stated, and compound it at the rate of 126.45 per cent. every ten years, and the result is largely over $3,000,000,000 in 1870, and over $7,000,000,000 in 1880, independent of the effect of any immigration succeeding 1860. If these results are astonishing, we must remember that immigration here is augmented population, and that it is population and labor that create wealth. Capital, indeed, is but the accumulation of labor. Immigration, then, from 1850 to 1860, added to our national products, a sum more than double our whole debt on the 1st of July last, and augmenting in a ratio much more rapid than its increase, and thus enabling us to bear the war expenses."

As the homestead privilege must largely increase immigration, and add especially to the cultivation of our soil, it will contribute vastly to increase our population, wealth, and power, and augment our revenues from duties and taxes.

As this domain is extended over fifteen States and all the Territories, the completion of these enlarged canals, embracing so large a portion of them, would be most advantageous to all, and the inducement to immigration would greatly increase, and immigration must soon flow in from Europe in an augmented volume. Indeed, when these facts are generally known in Europe, the desire of small renters, and of the working classes, to own a farm, and cultivate their own lands here, must bring thousands to our shores, even during the war. But it will be mainly when the rebellion shall have been crushed, the power of the Government vindicated, it authority fully re-established, and slavery extinguished, so as to make labor honorable everywhere throughout our country, and freedom universal, that this immigration will surge upon our shores. When we shall have maintained the Union unbroken against foreign and domestic enemies, and proved that a republic is as powerful in war as it is benign in peace, and especially that the *people* will rush to the ranks to crush even the most gigantic rebellion, and that they will not only bear arms, but taxes, for such a purpose, the prophets of evil, who have so often proclaimed our Government an *organized anarchy*, will lose their power to delude the people of Europe. And when that people learn the truth, and the vast privileges offered them by the Homestead Bill, there will be an exodus from Europe to our country, unprecedented since the discovery of America. The wounds inflicted by the war will then soon be healed, and European immigrants, cultivating here their own farms, and truly loyal to this free and paternal Government, from which they will have received this precious gift of a farm for each, will take the place of the rebels, who shall have fled the country.

We have seen that the total cost of our railroads and canals, up to this date, was $1,335,285,569, and I have estimated the probable cost of these enlarged works as not exceeding one-tenth of this sum, or $133,528,556. Let us now examine that question. We have seen that our 4,650 miles of canals cost $132,000,000, being $28,387 per mile, or less, by $8,395 per mile, than our railroads. It will be recollected that a large number of miles of these canals have already the requisite depth of seven feet, and width of seventy, and need only an enlargement of their locks. It appears, however, by the returns, that the Erie canal, the Grand Junction, Champlain canal, and the Black River, Chemung, Chenango, and Oswego, in all 528 miles, are all seven feet deep, and seventy feet wide, and cost $83,494 per mile, while the average cost of *all* our canals, varying from forty to seventy-five feet in width, and from four to ten in depth, was $28,387 per mile. Assuming $28,000 per mile as the average cost of the canals requiring enlargement, and $83,000 that of those per mile, having already the requisite dimensions, the difference would be $55,000 per mile, as the average cost of those needing increased dimensions.

The estimated cost, then, would stand as follows:

598 miles New York canals, enlargement of locks..................	$5,980,000
Enlarging dimensions, etc., of 1,696 miles, at $55,000 per mile	93,280,000
Total	$99,260,000

The conjectural estimate heretofore made by me was $133,528,556, or, one-tenth the cost of our existing railroads and canals, and exceeding, by $1,528,556, the cost of all our present 4,650 miles of canal. Deduct this from the above $133,528,556, leaves $34,268,556 to be applied to improving the St. Clair flats, the Mississippi river deepening its mouth, and for the ship-canal round the Falls of Niagara.

No estimate is now presented of the cost of the canal from Lake Champlain to the St. Lawrence, because that requires the co-operation of Canada.

The railroads of our country would increase their business, with our augmented wealth and population, especially in the transportation of passengers and merchandise. They would also obtain iron cheaper for rails, boilers and engines, timber for cars, breadstuffs and provisions for supplies, and coal or wood for their locomotives.

Great, however, as would be the effect of these works in augmenting our commerce, wealth and population, their results in consolidating and perpetuating our Union would be still more important. When the Ohio, Mississippi, and Missouri, and all their tributaries, arterializing the great valley, shall be united by the proposed routes with the Lakes, the St. Lawrence, the Hudson, Delaware, Susquehanna, Chesapeake, and Albemarle, what sacrilegious hand could be raised against such a Union? We should have no more rebellions. We should hear no more of North, South, East and West, for all would be linked together by a unity of commerce and interests. Our Union would become a social, moral, geographical, political and commercial necessity, and no State would risk by secession the benefit of participating in its commerce. We should be a homogeneous people, and slavery would disappear before the march of civilization, and of free schools, free labor, free soil, free lakes, rivers and canals. It is the absence of such a system (aided by slavery), *drawing the West and South-west, by a supposed superior attraction, to the Gulf*, that has led the South-west into this rebellion. But with slavery extinguished, with freedom strengthening labor's hand, with education elevating the soul and enlightening the understanding, and with such communications, uniting all our great lakes and rivers, East and West, all crowned with flourishing towns and mighty cities, with cultured fields and smiling harvests, exchanging their own products and fabrics, and those of the world, by flying cars and rushing steamers, revolt or disunion would be impossible. Strike down every barrier that separates the business of the North and East from that of the South and West, and you render dissolution impossible. In commerce we would be a unit, drawing to us by the irresistible attraction of interest, intercourse and trade, the whole valley of the Lakes and St. Lawrence. Whom God had united by geography, by race, by language, commerce, and interest, political institutions could not long keep asunder. Of all foreign nations, those which derive the greatest disadvantages from such an union would be England and France, the two governments which a wicked pro-slavery rebellion invites to attempt our destruction. With such a commerce, and with slavery extinguished, we would have the Union, not as it was, but as our fathers intended it should be, when they founded this great and free republic. We should attain the highest civilization, and enjoy the greatest happiness of which our race is capable. So long as slavery existed here, and we were divided into States cherishing and States abhorring the institution,—so long as free and forced labor were thus antagonized, we could scarcely be said to have a real Union, or to exist truly as a nation. Slavery loomed up like a black mountain, dividing us. Slavery kept us always on the verge of civil war, with hostility to liberty, education and progress, and menacing for half a century the life of the Republic. The question, then, was not, Will any measure, or any construction of the constitution, benefit the nation? but, Will it weaken or strengthen slavery? All that was good, or great, or national, was opposed by slavery—science, literature, the improvement of rivers and harbors, homesteads for the West, defenses and navies for the East. American ocean steamers were sacrificed to foreign subsidies, and all aid was refused to canals or railroads, including that to the Pacific, although essential to the national unity. Slavery was attempted to be forced on Kansas, first by violence and invasion, and then by fraud, and the *forgery of a constitution*. Defeated in Kansas by the voice of the people, slavery took the question from the people, and promulgated its last platform in 1860, by which all the Territories, nearly equal in area to the States, were to be subjected forever *as Territories* to slavery, although opposed by the overwhelming voice of their people. Slavery was nationalized, and freedom limited and circumscribed, with the evident intent soon to strangle it in all the States, and spread forced labor over the continent, from the North to Cape Horn. Failing in the election, slavery then assailed the vital principle of the republic, the rule of the majority, and inaugurated the rebellion. Slavery kept perjured traitors for months in the cabinet and the two Houses of Congress, to aid in the overthrow of the Government. Then was formed a constitution avowedly based on slavery, setting it up as an idol to be worshipped, and upon whose barbaric altars is now being poured out the sacrificial blood of freemen. But it will fail, for the curse of God and man is upon it. And when the rebellion is crushed, and slavery extinguished, we shall emerge from this contest strengthened, purified, exalted. We shall march to the step and music of a redeemed humanity, and a regenerated Union. We shall feel a new inspiration, and breathe an air in which slavery and every form of oppression must perish.

Standing upon these friendly shores, in a land which abolished slavery in the twelfth century, and surrounded by a people devoted to our welfare, looking westward, along the path of empire, across the Atlantic, to my own beloved country, these are my views of her glorious destiny, when the twin hydras of slavery and rebellion are crushed forever.

If our Irish adopted citizens could only hear, as I now do, the condemnation of slavery and of this revolt, by the Irish people; if they could hear them, as I do, quote the electric words of their renowned Curran against slavery, and in favor of *universal emancipation;* if they could listen, as they repeated the still bolder and scathing denunciations of their great orator, O'Connell, as he trampled on the dehumanizing system of chattel slavery, they would scorn the advice of the traitor leaders, who, under the false guise of democracy, but in hostility to all its principles, would now lure them by the syren cry of peace, into the destruction of the Union, which guards their rights, and protects their interests.

The Convention now assembled at Chicago, can do much to inaugurate a new era of civilization, freedom and progress, by aiding in giving to the nation these great interior routes of commerce and intercourse, in the centre of which your great city will hold the urn, as the long-divorced waters of the Lakes and the Mississippi are again commingled, and the Union linked together by the imperishable bonds of commerce, interest and affection.

With great respect, your obedient servant,

R J. WALKER.

LIST OF DELEGATES

TO THE

SHIP-CANAL CONVENTION.

MAINE.

Hon. Hannibal Hamlin	Bangor.	James H. Perley	Portland.
Ex-Gov. Israel Washburne	"	Daniel Wood	"
A. D. Manson	"	John W. Perkins	"
Phillip H. Brown	Portland.	Charles E. Jose	"
T. C. Hersey	"	C. F. Foster	"
Jedediah Jewett	"	Horatio N. Jose	"
John A. Poor	"	John A. S. Dana	"
John Lynch	"	John T. Gilman	"
James O. Brown	"	C. B. Stitson	"

NEW HAMPSHIRE.

Hon. T. M. Edwards................Keene.

VERMONT.

Lucius Tallett	Burlington.	F. F. Strong	Burlington.
John Dallen	"	William Early	Hartland.
L. B. Platt	"	M. M. Huntington	Rochester.

CONNECTICUT.

S. N. Gould	Cornel.	A. W. Barrows	Hartford.
R. R. Dake	Pittsfield.	C. C. Tiffany	New Haven.
Hon. Thomas F. Bogans	Conn.	John P. Elton	Waterbury.
J. C. Wheaton	"	Thomas Porter	"
F. H. Bissell	Hartford.	Thomas L. Gould	"

MASSACHUSETTS.

Hon. J. H. Clifford	New Bedford.	Daniel Pitman	Boston.
Dr. E. P. Able	"	J. S. Blackford	"
Ed. N. Howland	"	S. B. Stebbins	"
Daniel Sullivan	"	J. S. Edgerly	"
Dr. Lyman Bartlett	"	Robert Hill	"
H. W. Seabury	"	William G. E. Pope	"
Theodore Knowles	"	Gilman Currier	"
L. Bartlett	"	J. C. Converse	"
Samuel G. Bowden	Boston.	William Penson	"
Oliver Perkins	"	Daniel Shundy	"
Charles C. Fuller	"	Solomon Parsons	"
Hon. L. Sabine	"	L. M. Beardsly	"
Wm. Hilton	"	A. Howe	"

MASSACHUSETTS.

R. E. Damen	Boston.	E. Southworth	Springfield.
John Fulton	"	Dr. L. Hans	"
Charles E. Grant	"	J. E. Hood	"
Moses P. Grant	"	Hon. H. L. Dawes	North Adams.
William T. Battles	Worcester.	George E. Johnson	Brookfield.
J. G. Lyde	"	George H. Laflin	Pittsfield.
A. Smith	"	Abel Hull	Elder.
Ed. Barton	"	M. D. Southwick	Blackstone.
C. Triffin	"	S. G. Hubbard	Hartland.
Dr. John Green	"	P. S. Williams	Hoadley.
James Dunlap	Northampton.		

RHODE ISLAND.

S. Hutchins	Providence.	R. J. Arnold	Newport.
George C. Bonlow	Woonsocket.	C. Tiffany	"

NEW YORK.

Hon. Erastus Corning	Albany.	Joseph Candee	Buffalo.
William H. Read	"	Henry Mechell	"
G. M. Griffin	"	E. T. Proper	"
J. W. Holbrook	"	S. V. R. Watson	"
William B. Lewis	"	H. W. Rogers	"
Hon. Samuel B. Ruggles	New York.	S. Chamberlin	"
Gen. Hiram Walbridge	"	F. A. Alberger	"
Ex-Gov. John A. King	"	William Wilkinson	"
Alfred M. Hoyt	"	George Holt	"
James Packard	"	H. B. Wilder	"
Henry S. Downs	"	S. S. Gutherie	"
E. Masters	"	Richard Williams	"
William T. Blodget	"	P. L. Stenberge	"
Josiah M. Fisk	"	J. R. Bently	"
H. A. Tucker	"	G. R. Wilson	"
Dr. D. J. McGowen	"	C. C. F. Gay	"
Chauncy M. Du Pee	"	A. M. Clapp	"
James A. Hamilton	"	H. D. Dorr	"
Alexander McLead	"	E. P. Dorr	"
Dr. Julius Homberger	"	H. O. Corning	"
Edwin F. Arnold	"	J. Parker	"
G. Furman	"	Walter Johnson	"
Hon. Moses B. McClay	"	John Allen	"
R. M. Tittsworth	"	Daniel Bowen	"
C. D. Wallace	"	P. H. Bender	"
G. F. Wickes	"	David Gray	"
John Faitte	"	G. Himmenson	"
Hon. E. G. Spaulding	Buffalo.	Walter Young	"
E. P. Dow	"	E. R. Jewett	"
A. L. Griffin	"	J. M. Richmond	"
S. Lincoln	"	Henry Steiner	"
D. R. Williams	"	C. C. Wycoff	"
Dr. A. H. Harvey	"	E. G. Gray	"
E. Stork	"	James C. Evans	"
G. S. Hazard	"	Gilbert Molleson	Oswego.
Wilson Brunson	"	A. P. Grant	"
Henry W. Rogers	"	B. F. Greene	"
Wallace Johnson	"	Wills Nelson	"
Ed. Strock	"	J. Wendall	"
Ed. Tohill	"	Charles Wendall	"

NEW YORK — *Continued.*

J. C. Churchell	Oswego.
Joseph Owen	"
Leonard Ames	"
George Ames	"
F. A. G. B. Grant	"
L. B. Crocker	"
Hon. D. C. Littlejohn	"
Hon. Henry Fitzhugh	"
Henry Fitzhugh, Jr.	"
R. C. Mattoon	"
P. Burdy	"
H. L. Davis	"
D. McCallen	"
J. W. Petkin	"
Isaac Fellows	"
Daniel Lyons	"
William Averly	"
D. De Wolf	"
Frank Pulver	"
Willard Johnson	"
L. B. Robe	"
D. M. Callum	"
Daniel Lyons	"
M. J. Cummings	"
Hon. C. B. Sedgwick	Syracuse.
Charles Tallman	"
William H. Sparkland	"
Henry Baldwin	"
A. H. Hovey	"
Charles Andrews	"
Robert Graw	"
T. R. Porter	"
L. Gleason	"
W. H. Shackland	"
John P. Concade	Rochester.
Henry Way	"
A. Sherman	"
R. Sherman	"
W. Walker	"
S. M. Marsh	"
F. Vase	"
A. Vase	"
T. R. Porter	"
J. M. Winslow	"
J. R. Wilder	"
M. H. Mills	"
D. M. Denny	"
H. L. Denny	"
Charles Cook	Havana.
E. B. Holmes	Brockport.
Wm. Pearson	Orange.
B. F. French	Utica.
S. S. Faxon	"
F. C. Dowding	Attica.
H. J. Sickles	"
G. C. Baldwin	Troy.
E. C. Porter	"
J C. Hart	"
Thomas McManus	"
J. R. Prentice	"
R. D. Bradwell	Troy.
J. G. McMurry	"
S. B. Saxton	"
E. J. Hicks	"
Charles Eddy	"
A. B. Morgan	"
Charles A. Holmes	"
M. L. Tilly	"
Charles S. Richards	"
T. R. Sexton	"
F. A. Sheldon	"
William H. Young	"
John Sherry	"
F. C. Brinsmade	"
J. R. Trail	"
W. M. A. Nismal	"
James O. Leach	Balston Springs.
Jonathan S. Beach	"
James M. Badger	Brooklyn.
Charles Rowland	"
Rev. G. S. Corwin	Elba.
A. H. Hutchinson	Albion.
Henry J. Sickles	"
A. R. Patterson	"
S. C. Payne	"
V. V. Bullock	"
L. J. Peck	"
Jesse N. Seely	Pitcher.
E. A. Price	Palmyra.
G. W. Cuyler	"
Charles McClouth	"
Dr. James S. Whaley	Vernon.
William Wheeler	Deposit.
D. A. Ogden	Pen Yan.
John Fisk	Niagara.
J. N. Merchant.	Madison Co.
N. R. Walker	Livingston.
C. H. Mason	Monroe.
G. Chambers	Tompkins Co.
Judah Ellsworth	Saratoga.
D. Mariam	Whitehall.
M. F. Russell	Saugerties.
J. M. Boyes	"
B. Shaler	"
F. S. Laflin	"
S. G. Sowing	"
Griffin Smith	Lockport.
A. Frink	Warsaw.
Benjamin N. Huntington	Rome.
J. S. Sprague	Cooperstown.
David Wadsworth	Auburn.
Hiram W. Tuttle	Watertown.
Benjamin Haddock	"
Norman Sacknider	Ogdensburg.
J. W. Smith	"
D. M. Chapin	"
Henry G. Foot	"
W. C. Brown	"
David C. Judson	"

NEW JERSEY.

Hon. M. L. Ward........... Newark.
Joseph Jackson............ "
Dr. J. B. Sayers............Sussex.
John Blean................Mayesville.

PENNSYLVANIA.

Lucius Sneider.............Westport.
A. Reed..................Lewiston.
E. H. Mason ....Towanda.
Daniel Holmes.............Canton.
John R. Thomas...........Harrisburg.
Henry Palmer............. York.

OHIO.

Dr. E. Smith............... Dayton.
T. M. Wales........... .... Xenia.
C. L. Garron ....Cincinnati.
E. Conkling............... "
A. H. Agard................Sandusky.
John L. Johnson............ Toledo.
F. Bissell .... "
E. B. Hyde................ "
W. T. Walker.............. "
Henry Chase.............. "
W. W. Jones .... "
N. B. Morrison...........Highland Co.
M. B. Tuckins............. Oberlin.
H. E. Peck .... "
H. G. Little............... "
J. B. T. Marsh .... "
J. T. Irwin.................Crestline.
Philo Chamberlin ....Cleveland.
M. B. Scott............... "
C. N. Coe .... "
S. S. Coe .... "
R. J. Lyon.................Cleveland.
C. Bradford............... "
L. F. Vance .... "
W. Rockfellow .... "
R. K. Winslow .... "
G. N. Gardner .... "
J. H. Clark............... "
J. J. Weaver .... "
Philo Scovill.............. "
L. Case, Jr. .... "
J. E. Curry............... "
N. J. Boardman .... "
R. P. Spaulding .... "
A. G. Riddle............... "
A. N. Fairbanks........... "
S. D. Payne............... "
J. H. A. Borne............ "
Stilman Witt.............. "
J. N. Fitch................ "
R. Whitcare.............. Salem.

MICHIGAN.

Prof. A. B. Paliner........ Ann Arbor.
F. S. Stebbins............. "
R. J. Barry .... "
C. G. Clark, Jr. .... "
E. C. Seacmon .... "
J. M. Gregory............. "
Frank L. Stebbins .... "
J. B. Cupper..............Coldwater.
D. B. Dennis.............. "
S. B. Dennis............. "
L. D. Morris ....Ypsilanti.
B. Tullett................. "
E. S. Moore ....Three Rivers.
A. W. Ditmas............. "
J. A. Walter.............. "
E. A. Egery............... "
William Gould ....Kalamazoo.
R. May................... "
E. E. M. Little............. "
Hon. H. G. Wells .... "
M. L. Booth............... "
C. C. Comstock ....Grand Rapids.
Samuel Lewis ....Broadhead.
R. Johnson .... Adrian.
Morgan Johnson........... Detroit.
F. D. Hawley.............. "
A. E. Bissell .... "
Alex. Lewis............... "
J. D. Standish............. "
John W. Longyear.......... "
Albert Williams........... "
R. F. Johnson .... "
E. G. Merrick .... "
Wm. H. Craig............. "
John S. Patten .... "
Edmund Kanter .... "
Philo Parsons....... ...... "
R. W. King,............... "
James Aspinwall........... "
F. G. Bagley.............. "
T. H. Hinchman .... "
E. R. Mathers .... "
Smith Botsford .... "
B. G. Stimson............. "
Samuel Lewis .... "
Ray Haddock .... "
John G. Owen............. "
R. A. Beal................ "

MICHIHAN.

J. F. Conover .... Detroit.
E. J. Joyce .... "
F. Lambie .... "
John P. Wyckoff............ Pontiac.
E. W. Jenks................ Sturgis.
W. J. Baxter.............. Jonesville.
Charles Dickey ....Marshall.
Col. C. Dickey .... "
C. L. Miller .... Colon.
A. Clock ....Dowagiac.
W. J. Cornell .... Neoga.
D. O. Woodruff............. Niles.

INDIANA.

L. D. Glassbrook ....Sand Pier.
G. A. Hendricks.... .....South Bend.
William Miller .... "
George W. Julien..........Centreville.
E. L. Merrick .... Lake.
S. A. Freeman. ....Fort Wayne.
N. H. Buford............... "
S. Freeman .... "
A. H. Luther.... ....Valparaiso.
A. B. Judson ....St. Joseph.
H. E. Herbert .... "
Ed. Marshall ....Rolling Prairie.
A. B. Judson............Mishawaukee.
J. G. Wilson ....Charleston.
L. H. Hamlin ....Indianapolis.
M. Bishop ....Lafayette.

ILLINOIS.

Dr. Daniel Brainard .... Chicago.
J. W. Foster .... "
Frederick Tuttle............ "
G. G. Wolcott........... ... "
William Sturgis .... "
W. D. Houghteling .... "
J. A. Hahn .... "
C. G. Wicker .... "
J. T. Edwards..... .... "
H. E. Sargent.............. "
George M. Gray .... "
J. C. Fargo .... "
C. G. Hammond .... "
William B. Ogden .... "
George W. Gage............ "
D. H. Gage .... "
William A. Tucker.......... "
J. Q. Hoyt................ "
J. H. Kinzie............... "
Thomas Drummond..... ... "
Van H. Higgins .... "
P. W. Gates .... "
H. D. Colvin .... "
William Bross.............. "
J. W. Sheahan .... "
D. A. Cashman.............. "
J. L. Wilson................ "
S. J. Medill .... "
L. Brentano................ "
R. M. McChesney .... "
William McKindley .... "
John Wentworth............ "
John B. Preston............ "
Nathan H. Parker .... "
John N. Lynch .... "
R. M. Hough .... "
J. Medill .... "
J. S. Rumsey .... "
John L. Hancock .... "
W. E. Doggett .... Chicago.
John A. Nichols............ "
G. C. Cook .... "
H. W. Hinsdale .... "
H. A. Hurlbut.............. "
W. T. Shufeldt .... "
George Armour .... "
Clinton Briggs .... "
Henry W. King .... "
Nelson Tuttle .... "
James Robb .... "
I. N. Arnold .... "
F. C. Sherman. .... "
J. K. Botsford............. "
Merrill Ladd .... "
William Blair...... .... "
E. Burnham................ "
E. Hempstead .... "
Stephen Clary............. "
J. A. Walters .... "
L. J. S. Flint .... "
J. W. Smith .... "
Mark Skinner .... "
Walter S. Newberry .... "
Grant Goodrich .... "
Henry Farnam .... "
M. L. Sykes................ "
W. S. Gurnee .... "
Potter Palmer.............. "
Thomas B. Bryan .... "
J. B. Turner .... "
B. F. Ayer .... "
Phillip Conley. .... "
George H. Schneider........ "
H. J. Loomis .... "
E. M. Banker............... "
Charles Harmon...........Springfield.
William Day .... "
E. Rigney .... "

ILLINOIS — *Continued.*

N. J. Conkling	Springfield.
T. R. King	"
C. W. Mathews	"
John Armstrong	"
Vincent Ridgley	"
Sidney Pulsifer	Peoria.
H. Jewett	"
D. Hitchcock	"
J. P. Johnson	"
John W. Mansell	"
C. D. Rankin	"
A. G. Anderson	"
L Hurd	"
E. C. Ingersoll	"
J. C. Proctor	"
N. G. Millard	"
M. Locks	"
E. A. Andrews	"
Peter Sweet	"
Matthew Craig	"
Dr. R. Rouse	"
Benjamin Miller	"
Frederick Potter	"
M. W. McReynolds	"
H. Bronson	"
F. G. Warren	"
Lewis Hewil	"
Luther Card	"
John Weber	"
James Stewart	"
C. H. Smith	"
L. L. Guyer	"
Charles Farewell	"
David W. Cummings	"
George H. Catcel	"
R. Bills	"
A. J. Hodges	"
J. H. Thompson	"
W. Carrington	"
P. J. Matthews	"
E. Vemmun	"
A. M. Abbott	"
D. D. Snyder	"
G. W. Brayton	"
E. C. Ingersol	"
E. A. Andrews	"
R. Rouse	"
B. L. T. Bourland	"
C. C. Wood	"
Isaac Underhill	"
John H. Francis	"
John Waugh	"
F. Bohe	"
Henry Kuse	"
C. W. Rees	"
J. C. Procter	"
T. S. Bradley	"
F. G. Warren	"
J. F. Devin	"
W. A. Willard	Peoria.
P. C. Merwin	"
George W. Hitchcock	"
P. W. Dunne	"
P. R. K. Brotherson	"
N. Huggins	"
Wellington Loucks	"
Louis First	"
B. K. Harrington	"
Anthony Rink	"
Matt Goodsfield	Ottawa.
B. N. Purcell	"
J. E. Baldwin	"
W. C. Richardson	"
John Arnold	"
Washington Bushnell	"
William Hickling	"
J. V. A. Hoes	"
John C. Champlin	"
B. C. Cook	"
S. W. Cheaver	"
J. F. Nash	"
William Reddick	"
Ed. Eames	"
David Walker	"
B. M. Boyle	"
P. K. Lealand	"
John Manley	"
F. C. Gibson	"
S. W. Raymond	"
S. Crook	"
John Hossack	"
J. E. Porter	"
Dr. C. Hurd	"
Gen. J. N. Singleton	Quincy.
Judge O. C. Skinner	"
C. A. Jayne	"
U. S. Penfield	"
Rev. M. N. Willis	"
William N. Anso	"
John Melby	"
David J. Jameson	"
M McVay	"
William Minkleman	"
John Locke	"
C Kathman	"
William Mitchell	La Salle.
Charles Rosenburg	"
C. C. Vattis	"
John Vongasser	"
F. G. Crane	"
J. G. Hatch	"
Ed. Mason	"
O. N. Adams	"
Phillip Conlin	"
Thomas Floyd	"
Dr. L. B. Larkin	"
Isaac Hardy	"
E. B. Mason	"

ILLINOIS — *Continued.*

Joseph Bliss	La Salle.
C. S. Rosenburg	"
S. Sullinger	"
Dr. Kelsey	"
B. F. De Merriett	"
L. B. Larkin	"
F. J. Craine	"
Col. D. Taylor	"
J. S. Thrasher	"
Solomon Slager	"
L. Ebill	"
R. Cady	"
J. Hardy	"
N. C. Cannon	"
D. McKey	"
V. G. Hatch	"
John Collins	"
Joseph Blush	"
G. N. Teal	"
L. C. Hathaway	"
B. T. Filts	"
L. H. Emens	"
N. B. Bristol	"
S. E. Muer	"
H. A. Buttler	"
W. H. Brown	"
M. D. Calkins	"
James H. Dicken	"
H. Mayo	"
C. H. Joyce	"
David Stung	"
E. Vander Fields	"
J. D. Hempstead	"
A. W. Magill	"
John Hoosick	"
William A. Punnell	Peru.
C. W. Brown	"
James Bartin	"
Dr. S. G. Smith	"
J. B. Champney	"
William D. Parsons	"
C. W. Munger	"
A. W. White	"
George D. Ladd	"
David Lininger	"
Samuel N. Troth	"
A. Sapp	"
Paul Bohemue	"
William Waker	"
H. Zimmerman	"
John Aaron	"
William Rouch	"
N. Sherwood	"
J. F. Evans	"
N. Evans	"
E. S. Holbrook	"
F. Benkhart	"
M. Rush	"
J. J. Wicks	"
J. Beaumont	Peru.
James Barton	"
Wm. B. Day	"
G. D. Ladd	"
Isaac Abrahams	"
John Brown	"
R. C. Kelly	"
J. P. Hadigan	"
E. S. Winslow	"
Peter Barnes	"
William Paul	"
J. Kaiser	"
Jo. Koons	"
P. K. Bechard	"
G. S. Eldridge	"
Benjamin Ream	"
J. S. McCormick	"
William Chumas	"
H. Silver	"
C. Charles	"
C. N. Munger	"
L. Runnell	"
S. N. Harding	"
J. F. Weeks	"
William Walker	"
John Brown	"
William B. Day	"
D. C. Young	Joliet.
John N. Hill	"
M. C. Russell	"
W. B. Hanley	"
W. W. Stephens	"
W. J. Haines	"
D. Cole	"
C. Sams	"
B. U. Sharp	"
A. H. Day	"
F. L. Cagwin	"
F. Goodspeed	"
W. J. Barrett	"
E. Savage	"
C. E. Ward	"
W. H. Mosher	"
J. Kerchinial	"
J. Millar	"
W. H. Powell	"
C. E. Van Aiken	"
N. C. Wood	"
E. V. Bronson	"
H. Bates	"
R. Rhodes	"
James Curtis	"
William Curtis	"
Dr. Alden	"
C. E. Munger	"
H. Lowe	"
Moses Hall	"
K. T. Hammond	"
G. W. Cassidy	"

ILLINOIS — *Continued.*

W. F. Barrett	Joliet.
H. Hook	"
Sheriff Monroe	"
K. Doolittle	"
James Beamont	"
O. Fox	"
J. Boons	"
H. Larraway	"
J. Hogan	"
F. Mack	"
A. McIntosh	"
J. L. Braden	"
J. McRoberts	"
L. Ziph	"
W. R. Steele	"
J. C. Breckenridge	"
J. S. Morgan	"
N. E. Wagner	"
C. Morrill	"
A. Leach	"
W. H. Carlan	"
L. P. Sunger	"
D. Cox	"
W. W. Stevens	"
U. Osgood	"
M. K. Sebastine	"
Nelson Fay	"
H. S. Carpenter	"
L. N. Bates	"
N. Danforth	"
Jacob L. Bower	"
J. McRoberts	"
William Grenton	"
William Grenton, Jr.	"
James Johmot	"
R. D. Brown	"
A. H. Taylor	"
C. Chapman	"
T. O'Brien	"
N. Gagan	"
E. B. Harpham	"
George N. Walker	"
R. R. Simons	"
Francis Lowe	"
T. R. Hunter	"
M. L. Adams	"
W. J. Harris	"
Rodney House	"
J. Laraway	"
George O. Curtiss	"
G. W. Cassidy	"
T. L. Breckenridge	"
J. G. Brown	Polo.
D. J. Pinckney	"
John W. Hitt	"
R. J. Hitt	"
A. M. Hitt	"
A. Sanborn	"

Norman Hanks	Polo.
Hamilton Newton	"
Henry Burlington	"
Francis Strunrauch	"
R. Wagener	"
C. K. Williams	"
Norman Hawks	"
C. K. Williams	"
George Yates	"
D. Z. Herl	"
C. S. Ritz	"
Ruben Wagner	"
George Weaver	"
James G. Brown	"
J. M. Cobleigh	Fulton.
O. B. Crosby	"
M. Robbins	"
Frank Clendenin	"
D. S. Spafford	"
E. B. Warner	"
S. Strawder	"
A. Farrington	"
John McDonald	"
R. Sage	"
R. Beswick	"
F. Sackett	"
L. K. Hawthorne	"
S. S. Patterson	"
Col. Kilgore	"
Charles Wright	"
W. Loomis	"
S. Sampson	"
B. Arey	"
L. E. Dodge	"
W. C. Snyder	"
J. McCoy	"
L. McCartney	"
J. H. Perry	"
J. Rice	"
E. B. Wells	"
C. Pease	"
J. Phelps	"
Oliver Baker	"
A. J. Warner	"
P. B. Brown	"
A. J. Fuller	"
L. H. Robinson	"
J. Cobleigh	"
S. H. McCrea	"
J. E. Duffin	"
R. E. Logan	"
Leander Smith	"
B. Robinson	"
H. C. Fellows	"
H. W. Fonter	"
Hon. D. E. Dodge	"
W. S. Barnes	"
D. Efner	"

ILLINOIS — *Continued.*

Name	Residence
A. J. Matteson	Fulton.
W. H. Whipple	"
A. S. Ferguson	"
William Prothrow	"
J. B. Emmons	"
J. Marshall	"
J. T. Atkinson	"
M. Mead	"
R. H. Jenkins	"
A. Morse	"
L. Rice	"
J. G. Banes	"
D. Cleveland	"
C. W. Alysworth	"
F. E. Marcellus	"
O. Cowles	"
P. D. Wright	Fulton City.
W. T. Wiley	"
G. B. Raymond	Elgin.
N. G. Hubbard	"
O. Davidson	"
D. F. Barclay	"
John Wait	"
George Hassen	"
G. P. Harvey	"
A. P. Fish	"
William Panton	"
E. W. Kimball	"
Ira Cockley	"
B. F. St. Clair	"
John Ransted	"
M. Mallory	"
J. Teeft	"
J. Salisbury	"
R. W. Padelford	"
W. Floyd	"
John Kiser	"
Henry Sherman	"
J. M. Truesdale	"
C. B. Holly	"
A. J. Joslyn	"
Thomas Bruton	"
E. W. Cook	"
M. C. Traver	"
M. W. Elthrop	"
B. F. Truesdale	Amboy.
W. C. Ives	"
C. D. Vaughan	"
J. L. Baker	"
R. T. Adams	"
J. S. Briggs	"
F. R. Dutcher	"
William S. Cartrall	"
N. J. Swartwood	"
Almon Ives	"
Josiah Little	"
N. L. Chase	"
Hiram Nance	Kewanee.
James Goodspeed	Kewanee.
Dr. T. D. Fitch	"
William Albro	"
H. H. Bryan	"
E. P. Johnson	"
Theodore Nyley	"
S. S. Andrews	"
H. B. Bryant	"
George C. Howe	"
George F. Atwater	"
Nelson Lee	"
J. Hopkins	"
E. P. Johnson	"
H. T. West	"
G. C. Hurd	"
G. Atwater	"
F. Gentom	"
J. A. Lyon	"
H. S. Lay	"
E. C. Stone	"
J. R. Preston	"
H. C. Parker	"
H. T. West	"
George L. Herrick	Dixon.
Joseph Ball	"
George P. Goodwin	"
James K. Edsall	"
William Godfrey	"
A. L. Porter	"
Joseph Crawford	"
S. S. Williams	"
B. B. Higgins	"
B. F. Shaw	"
N. Underwood	"
W. T. House	"
Harvey Morgan	"
James McKenney	"
Samuel C. Ellis	"
Q. Ely	"
A. W. Pitts	"
Z. H. Luckey	"
James Ulley	"
Dr. O. Everitt	"
Hon. W. W. Heaton	"
Isaac Jones	"
Col. S. Noble	"
A. K. Norris	"
James Fletcher	"
A. McFerren	"
P. M. Alexander	"
Josiah Little	"
E. W. Dutcher	"
Martin Wright	"
F. H. Luckey	"
T. Wood	"
Joseph Utley	"
J. L. Camp	"
W. Holton	"

ILLINOIS — *Continued.*

Name	Place
A. Brown	Dixon.
H. B. Foulk	"
William Uhl	"
William M. Clark	"
E. Wood	"
D. P. Hall	"
Dan Castle	"
E. S. Fargo	"
George H. Page	"
Hon. John Van Estace	"
A. E. Darling	"
O. M. Alexander	"
L. Wood	"
J. B. Williams	"
G. R. Fulton	"
John Williams	"
H. B. Hills	"
P. J. Wood	"
Thomas Lowe, Jr	"
L. M. West	Rockford.
Anson Miller	"
E. C. Dougherty	"
J. H. Grow	"
C. G. Manlin	"
R. Betty	"
A. Hammond	"
D. P. Trupant	"
Peter Mabie	"
R. Jackson	"
L. M. Owen	"
N. B. Thompson	"
J. G. Penfield	"
J. C. Brown	"
W. B. Webb	"
W. W. Wood	"
W. G. Rea	"
E. Kirk	"
A. B. Skinner	"
E. H. Lansing	"
C. A. Shaw	"
A. S. Miller	"
T. B. Robertson	"
L. M. West	"
R. P. Lane	"
W. B. Slaughter	"
Thomas Kerr	"
William Lathrop	"
C. J. Horsman	"
H. J. Sawyer	"
O. A. Pennoyer	"
J. D. Sanders	"
C. W. Murfeldt	"
Morey Brown	"
George Lincoln	"
H. D. Frost	"
H. P. Kimball	"
C. G. Manland	"
R. D. Hill	"
Benjamin Wingate	"
C. C. Briggs	Rockford.
J. Sanders	"
William Talcott	"
Charles Williams	"
W. Wingate	"
Thomas Dement	"
G. A. Sanford	"
E. H. Baker	"
J. G. Manlove	"
B. E. Wingate	"
G. H. Dennett	"
R. A. Sanford	"
F. G. Ferguson	"
S. M. Church	"
D. Wallach	"
F. Blackman	"
G. W. Knox	"
O. Adams	"
D. S. Penfield	"
C. Breastead	"
T. S. Day	"
E. Day	"
H. P. Kimball	"
A. Hammond	"
R. Jackson	"
M. Brown	"
C. M. Mentfelt	"
William B. Stoughter	"
Jacob Locke	Princeton.
J. R. Earnest	"
C. C. Lattemier	"
W. O. Chamberlin	"
Charles Baldwin	"
J. Mercer	"
C. Jones	"
D. Robinson, Jr.	"
John G. Bradley	"
Charles S. Kelsey	"
M. T. Peters	"
C. P. Allen	"
George M. Radcliff	"
S. M. Dunbar	"
C. P. Allen	"
N. Matteson	"
Dr. C. H. Yager	"
L. H. Yarwood	"
Hon. Owen Lovejoy	"
A. Swantz	"
S. M. Knox	"
A. T. Colton	"
John B. Bubach	"
J. J. Carpenter	"
A. A. Shannon	"
L. J. Colton	"
Joshua R. Brown	"
Silas Butty	"
P. Fayercrants	"
Henry C. Reed	"
Milo Kendall	"

ILLINOIS — *Continued.*

Name	Place
George S. Paddock	Princeton.
J. V. Thompson	"
C. Barni	"
J. S. Miller	"
John Mansfield	"
John Bryant	"
Benjamin Newell	"
J. H. Pattison	"
John Brady	"
G. H. Phelps	"
A. Swanzy	"
L. S. Clossen	"
L. S. Smith	"
R. J. Templeton	"
N. Mattoon	"
W. S. Gale	Galesburg.
E. F. Thomas	"
C. J. Fenton	"
A. Mattison	"
W. A. Wood	"
C. S. Colton	"
Edwin Post	"
George W. Wood	"
Isaac Merrill	"
A. T. Boon	"
Leonard Miller	"
Alfred Knowles	"
W. B. Lee Barron	"
H. N. Keightly	"
H. W. Washburn	"
J. D. Hand	"
O. S. Pitcher	"
F. S. Short	"
L. C. Field	"
E. F. Davis	"
E. M. Cook	"
F. G. Harris	"
John A. Marshall	"
Chauncy Adams	"
M. D. Cook	"
B. O. Carr	"
F. R. Bartlett	"
C. H. Mathews	"
A. W. Martin	"
George J. Bergen	"
F. J. Hale	"
John C. Stuert	"
W. B. Patterson	"
G. M. Hunt	"
Warren Willard	"
Lozal C. Field	"
A. Merrell	Aurora.
S. McCarthy	"
C H. Goodwin	"
D. P. Waterman	"
John Holbrook	"
A. C. Gibson	"
N. H. Hawkins	"
G. W. Querean	"
S. R. Wagner	Aurora.
C. C. Earle	"
Samuel Hoyle	"
Thomas Judd	"
James Robinson	"
E. R. Allen	"
S. A. Winslow	"
J. G. Stolp	"
W. J. Strong	"
B. H. Hackney	"
Austin Mann	"
L. W. Grey	"
E. Canfield	"
P. A. Allire	"
L. D. Brady	"
John Rising	"
E. Gillett	"
W. H. Miller	"
A. R. Palmer	"
B. B. Waterman	"
W. V. Phum	"
James G. Barr	"
Charles Wheaton	"
W. C. Watkins	Bloomington
C. R. Overman	"
C. A. Haner	"
S. L. Lord	"
J. L. Rice	"
A. J. Merriman	"
H. P. Merriman	"
K. H. Well	"
J. F. Mund	"
A. Elden	"
N. Dixon	"
D. D. Haggard	"
Charles S. Jones	"
N. R. Smith	"
N. H. Stennett	"
C. Wakefield	"
C. R. Park	"
A. B. Ives	"
William Perry	"
H. Noble	"
S. N. Noble	"
John Manica	"
J. C. Baker	"
J. L. Rice	"
William Perry	"
A. C. Markham	"
James S. Ewing	"
A. E. Stephenson	"
Charles F. Ingalls	Lee Co
H. E. Williams	"
William M. Van Epps	"
Josiah Little	"
Lewis Clapp	"
Amos C. Stedman	"
Alexander Brown	"
N. Holton	"

ILLINOIS — *Continued.*

Name	Place
E. B. Stiles	Lee Co.
A. M. Chadwick	"
N. H. Gardner	"
G. W. Braton	"
T. C. Angier	"
P. M. Alexandre	"
William Buttler	"
George M. Deland	"
David Utley	"
Lewis Merreman	"
Benjamin Brooks	"
James L. Camp	"
David Holly	"
John Dement	"
James L. Hanley	"
A. Cogswell	"
E. M. Dutcher	"
A. E. Wilcox	"
A. D. Moon	"
Alexander Charters	"
Charles Dement	"
James Moore	"
Isaac Sietz	"
Andrew McPherson	"
Isaac Jones	"
John H. Cropsey	"
W. H. House	"
H. B. Tinker	"
Samuel C. Ellis	"
James Fletcher	"
Joseph Utley	"
G. E. Haskel	"
Charles Brackett	"
Charles Curby	"
William E. Ives	"
John C. Jacobs	"
Enoch Wood	"
W. H. Hanson	"
Fred. Coe	"
B. H. Truesdale	"
D. J. Harris	"
A. R. Whitney	"
C. J. Reynolds	"
William Uhl	"
John Beale	"
Wallace Judd	"
M. O. Woodford	"
James Crawford	"
Charles Godfrey	"
George E. Haskill	"
James Godfrey	"
William H. Godfrey	"
O. Emitt	"
A. J. Brubaker	"
L. Wood	"
J. B. Williams	"
John Williams	"
Martin Wright	"
Samuel J. Butler	"
G. B. Reynolds	Lee Co.
W. A. Riley	"
S. Wood	"
M. S. Wood	"
Lewis Clapp	"
A. Gavins	Geneseo.
S. C. Campbell	"
W. P. Cook	"
C. A. Fisher	"
A. Crawford	"
J. M. Hasford	"
J. J. Town	"
Dr. J. S. Shaw	"
James Bernard	"
E. M. Stewart	"
J. B. Jones	"
Dr. S. J. Hume	"
O. J. Stough	"
J. O. Hood	"
Alexander White	"
William Smith	"
A. Tuttle	"
J. Bungesor	"
W. B. Boat	"
J. S. Hutchins	"
J. W. Wilson	"
A. W. Perry	"
N. B Gold	"
Charles Gill	Union.
W. Carr	"
William H. Alden	"
J. Preston	"
J. G. Botsford	"
T. Henry	"
W. P. Abbott	"
F. C. Filmore	"
Marcus White	"
O. S. Jenks	"
R. M. Patrick	"
E. F. McKenney	"
H. Prathe	Decatur.
J. G. Taylor	"
E. D. Smith	"
A. W. Hardy	"
A. J. Crain	"
D. L. Allen	"
A. A. Murry	"
E. A. Barnwell	"
T. K. Edmundson	Clinton.
James C. Wicker	"
G. Stout	Blackberry Station.
N. N. Kendall	"
Elma Kendall	"
J. P. Bartlett	"
G. N. Knight	Blackberry.
J. W. Foss	"
C. Reed	"
H. L. Rapelge	Kaneville.
Joseph Knapp	Orion.

ILLINOIS — *Continued.*

W. H. Johnson ....New Milford.
D. S. Pardee .... "
G. B. Fletcher .... Ogle.
John Williams .... "
P. B. Brown ....Morrison.
W. H. Gregory .... "
D. F. Spafford............. "
Frank Clevidan............ "
John E. Duffien .... "
N. Thompson .... "
A. J. Warner .... "
A. J. Fuller............... "
A. J. Matteson ... . "
O. W. Gage............... "
William Hill .... "
A. Ten Eyck ....Irvington.
William Newell .... Paris.
A. Z. Frogdon............. "
John Adams .... Berlin.
E. S. Richardson...........Chebanse.
Joseph Heigh .... "
Charles Bard.............. "
William G. Gage............ Detroit.
A. C. Jackson ....Whiteside Co.
A. K. Wheeler .... Kendall.
Gen. Thomas Maehn......... Albany.
E. K. Lester............... Monroe.
Albert Field, Jr............ "
C. H. Phillips ....Desplaines.
A. F. Miner............ "
John P. Taylor ....Lafayette.
Lewis Payne .... "
James Taylor .... "
F. Chase .... "
O. P. Hathaway...........Marsailles.
J. T. Nichols.............. "
H. B. Patrick............Bloomingdale.
C. H. Meacham........... "
W. G. Waterman ....Barrington.
G. P. Hardy............... Shelby.
Charles Sprague........... Cass Co.
William Cunningham .... "
E. B. Leonard .... "
George Plane .... "
William Chase .... "
H. C. Chansey .... "
J. O'Neill .... "
Charles Elburgh........... "
Henry Seger .... "
J. A. Avery .... "
Charles D. Moody .... "
Henry J. Foster .... "
Willis M. Hitt .... "
Lewis Ellsworth .... "
A. B. McConnell............ "
James N. Brown............ "
J. H. Pickeral .... "
A. J. Matteson .... "

H. S. Osborne .... Cass Co.
William Kile .... "
Horace Bullings .... "
Isaac Overall.............. "
J. C. Leonard .... "
R. Bigsby.................Marengo.
D. W. Lawrence........... "
J. H. Bagley .... "
E. H. Seward.............. "
L. M. Stevens .... "
F. Safford................. "
T. R. Ircenbrack........... "
A. B. Coon .... "
E. F. McKenna............ "
G. Hackley .... "
L. M. Hart................ "
O. Hager .... "
C. C. Miller............... "
C. H. Hibbard............. "
A. D. Kulo................ "
D. Bixton................. "
W. A. Treat .... "
Marcus White............ "
R. Brigsby................ "
D. E. Haltman .... "
C. Lancing................ "
David Teed .... "
Peter W. Duty .... "
Alden Jewett .... "
G. Y. Smith... ....Plainfield.
J. J. Touser............... "
G. W. Touser .... "
M. Dice .... "
A. T. Dice................ "
W. J. Hayes............... "
John Hays .... "
J. W. McBride .... "
J. Hager .... "
E. M. Pike .... Morris.
S. S. Benton .... "
George F. Barstow.......... "
Phillip Collins............. "
S. C. Collins .... "
C. Grant.................. "
Robert Longworth.......... "
William White............ "
D. R. Miller............... "
John B. Davidson .... "
George Gallaway .... "
George Fisher............. "
E. W. Hubbard............ "
J. W. Lawrence .... "
C. Comerford .... "
C. E. Sutherd .... "
Henry Clapp .... "
Joseph Hicks .... "
Edward S. Williams .... "
W. R. Conklin .... "

ILLINOIS — *Continued.*

Name	Place
Perry A. Armstrong	Morris.
S. W. Harris..............	“
J. N. Reading.............	“
C. U. Gould...............	“
L. W. Claypool............	“
J. Dicky..................	“
H. M. Hinkley.............	“
G. F. Brown...............	“
E. Hanna..................	“
B. C. Church..............	“
Benjamin Ollin............	“
Luke Hall.................	“
James A. Doane............	“
O. B. Galusha.............	“
L. J. Russell.............	Lacon.
H. S. Crane...............	“
G. W. Thompson............	“
Henry Steiner.............	“
N. D. Ford................	“
A. N. Ford................	“
E. Kaldenbaugh............	“
D. B. Weir................	“
Henry Sherburn............	“
A. Garrett................	“
J. S. Thompson............	“
James B. Martin...........	“
John A. Purley............	“
J. C. Granger.............	“
J. R. Taggett.............	“
L. J. Russell.............	“
W. E. Hancock.............	“
J. W. Hancock.............	“
Samuel Pomroy.............	“
J. B. Forbes..............	“
M. Burry..................	“
B. Waitzell...............	“
John Piper................	“
Joseph W. Wilder..........	“
H. D. Craye...............	“
W. M. Clarkson............	Henry.
R. Hodgman................	“
P. M. Jenny...............	“
William Fountain..........	“
C. M. Baker...............	“
A. M. Poole...............	“
J. H. Jones...............	“
J. M. Purple..............	“
J. C. McCurdy.............	“
E. H. Hutchins............	“
George Scott..............	“
Eli Smith.................	“
A. Catlin.................	“
R. Clark..................	“
Samuel Parker.............	“
A. N. Dickinson...........	“
C. S. Woodward............	“
John Roberts..............	“
D. C. Hull................	“
John Coch.................	“
John Kline................	Henry.
A. Hoaglan................	“
Lewis Schneider...........	“
Albert Ramsey.............	“
Enoch Sayer...............	“
James Rodgers.............	“
J. T. Thornton............	“
John Koch.................	“
J. F. Son.................	“
B. Zaker..................	“
Hon. Lyman Trumbull.......	Alton.
S. Maxwell................	Livingston Co.
J. R. Walganot............	“
J. Whetmore...............	“
Dr. S. Wood...............	“
J. V. Taylor..............	“
H. Eldridge...............	“
J. A. Vincent.............	“
Rev. B. F. Scriven........	“
E. Tracy..................	“
N. S. Stackpole...........	“
B. F. Blackwell...........	“
A. E. Harding.............	“
J. W. Sterell.............	“
J. E. Dye.................	“
L. E. Kent................	“
William Perry.............	“
M. A. Collins.............	“
N. C. Babcock.............	“
S. L. Fleming.............	“
Hon. M. B. Patty..........	“
Hon. R. P. Morgan.........	“
Hon. J. Angle.............	“
S. Strong.................	“
Dr. J. H. Hagerty.........	“
S. T. K. Prime............	“
L. Kinyon.................	“
J. Little.................	“
Dr. J. Croswell...........	“
H. Rumington..............	“
H. M. Gilbert.............	“
N. R. Walker..............	Livingston.
Stephen Dunlap............	Jacksonville.
J. Burden.................	“
William Bronson...........	“
M. Payers.................	“
Andrew McFarland..........	“
H. Bendan.................	“
J. L. Baily...............	“
J. T. Alexander...........	“
J. Duncan.................	“
William Brown.............	“
Stephen Fulton............	“
David Prince..............	“
William Thomas............	“
H. B. McClure.............	“
J. W. King................	“
C. S. Goltray.............	“
A. McDonald...............	“

ILLINOIS — *Continued.*

George M. Chambers......Jacksonville.
Jonathan Nedey............ "
E. Greenleaf............... "
M. McConnell.............. "
Edwin Scott............... "
William Hamilton, Jr........ "
John Hewey............... "
John Trebeau.............. "
John Henry............... "
J. Duncan................. "
J. C. Southwick............Waukegan.
M. C. Form............... "
L. H. Brayman............. "
J. P. Nichols.............. "
N. J. Sans................. "
J. H. Flinn................ "
Hiram Hoagan.............. "
T. Winter................. "
Rev. E. P. Wright.......... "
C. M. Worthington.......... Shelby.
V. J. Adams............... "
John Klive................ "
J. Hunt................... "
J. V. McKinney............ "
J. H. Tuttle............... "
B. S. Weaver.............. "
W. M. Kilgore............. "
Ralph Sage................ "
W. McCane................ "
John Wood................ "
Henry Moore.............. "
L. B. Weatherby........... "
J. C. Mickel.............. "
L. K. Hawthorn............ "
B. C. Cablitz............. "
B. B. Stoddard............ "
Wm. A. Sanborn........... "
F. Sackett................ "
J. E. McPherson........... "
John Shephard............ "
Thomas A. Galt........... "
F. Sackett................ "
F. B Hubbard............. "
William Cook............. "
L. Hapgood............... "
Jacob B. Black............. Bristol.
Reuben Hunt.............. "
James Boomer............. "
John A. Cook.............. "
H. F. Day................. "
Charles Chandler........Chandlerville.
H. C. Robbins.............. Newark.
Albert Cook............... "
James Washburn........... "
G. W. Winchell............ "
John Thomburgh..........Channahon.
N. Smith.............Babcock's Grove.
W. B. Brown.............. Oneida.

H. P. Griswold............Hancock Co.
L. Q. Hewins.............. "
J. Y. Bullard.............. "
J. Huestis................ "
A. Goodell................ "
W. A. Lockie.............. "
O. F. Rowley.............. "
J. W. Semple..............Monmouth.
William Pressely.......... "
John Langdon............. "
Albert Mitchell............ "
Judge John Porter......... "
S. Wood.................. "
William F. Smith.......... "
E. C. Harrington............ Lodi.
S. Lewis.................. "
E. P. Raperston........... "
J. Shaver................ Lawrence.
F. White................. La Moille.
David Fletcher............ "
A. E. Blodgett............ "
Samuel Edwards........... "
Henry Snyder.............Berrien Co.
James Scott............... Oquawka.
Hiram Rase............... "
Lewis Leslie.............. "
C. M. Harris.............. "
E. C. Thomas.............Woodstock.
J. L. Trowbridge........... "
J. C. Trowbridge.......... "
Hon. T. D. Murphy......... "
J. A. Parish.............. "
P. G. Schriever............ "
E. E. Thomas.............. "
M. Johnson............... "
J. H. Slaum............... "
J. H. Johnson............. "
John Donnelly............. "
Neil Donnelly............. "
M. W. Hunt............... "
A. W. Tappen.............. "
J. C. Hetzel............... Dwight.
H. Elkridge............... "
Richard P. Morgan......... "
Thomas Little............. "
C. S. Newell.............. "
Alfred Castle..............Wyoming.
E. M. Small................ Galena.
James M. Spratt........... "
C. R. Pickens............. "
C. S. Stephenson.......... "
J. R. Maynor.............. "
A. S. Chitlin.............. "
C. A. Perkins............. "
James Rood............... "
B. E. O. Dill.............. "
A. S. Brink............... "
J. S. Wagener............. "

ILLINOIS — *Continued.*

Joseph Morsley .... Geneva.
E. Peck .... "
B. DeWolf................. "
F. H. Bowman.............. "
G. W. Hall .... Mendota.
Nelson Dean...............Sandwich.
Charles Kennecott .... "
J. H. Culver .... "
George H. Cox............. "
J. E. Phelps.............. "
J. H. Carr................ "
A. Adams .... "
A. A. Feen .... "
J. C. Pettitt............. "
George Dean...............Belvidere.
John H. Herbert .... "
M. M. Boyce .... "
Alexander Neely .... "
George W. Munch........... "
O. H. Wright .... "
J. S. Hildrup............. "
Isaac R. Mudge............ "
S. R. Fox................. "
J. D. Wilson.............. "
William Corning........... "
J. D. Stowe............... "
William F. Hovey .... "
W. R. Dodge............... "
John Yourt .... "
W. H. Gilman.............. "
D. T. Pliney .... "
J. N. Brockway .... "
A. Burdick ....Lockport.
William Gooding........... "
John B. Preston........... "
G. A. Parks............... "
E. C. Fellows............. "
N. S. Rafferty .... "
James Dox................. "
F. G. Harris .... "
W. F. Ireson .... "
Thomas Norton............. "
David T. Louis .... "
H. S. Mason .... "
S. A. Snits .... "
T. M. Fish................ "
G. Williams............... "
J. G. Wright ....Naperville.
D. J. Brown .... "
Andrew Faudly .... "
Lewis Ellsworth .... "
Z. C. Brayton ....Kankakee.
Lyman Wooster .... "
B. T. Clark .... "
C. S. Wiltzee .... "
Jonathan Crews .... "
H. M. Kiser............... "
D. G. Bean .... "

John P. Gamble............Kankakee.
Emory Cobb .... "
C. C. Wilcox.............. "
D. S. Parker .... "
G. H. Andrews............. "
H. Whitmore .... "
W. G. Swanell .... "
G. V. Hewling............. "
L. Milk................... "
John Siles................ "
A. S. Vail ....Kankakee City.
E. Curtis................. "
J. B. Hamilton .... "
J. B. Worcester .... "
Elder Higbee.............. "
Edwin R. Beardsly......... "
Samuel Hawley .... "
E. B. Warriner............ "
John Dunham............... "
Thomas Kerr............... "
John Castle............... "
R. N. Murry .... "
F. Worcester.............. "
W. H. Patterson........... "
J. M. Perry .... "
Henry Boice, Jr........... "
Charles R. Starr.......... "
A. Bartlett............... "
James McGrew.............. "
N. G. Halsey.............. "
T. P. Bonfield............ "
R. Beebe .... "
S. R. Moore............... "
J. M. Wood................ "
J. A. Brown............... "
Nelson Duggan............Putnam Co.
C. H. Whitiker............ "
J. C. Leonard ....Beardstown.
C. D. Hobbitt ....St. Augustine.
J. C. Thompson .... Canton.
J. G. Piper .... "
M. Overman................. "
S. Y. Thornton .... "
L. C. Maynard .... "
S. Guttentag .... "
R. W. Dewey .... "
Thomson Smith .... Dement.
John H. Rowe............... "
B. W. Herring .... "
John V. Gale............... "
A. Barnum .... "
N. Stomgley .... "
Thomas Smith .... "
A. Diamon.................. "
Galva Welworth ....Courtland Station.
William Merrick........... Orland.
J. H. Merrick .... "
P. H. Thompkins .... El Paso.

ILLINOIS — *Continued.*

J. B. Fulton................Fairbury.
L. M. Marsh................ "
J. F. Blackburn .... "
H. L. Marsh .... "
James L. Alieson............ Mattoon.
B. Pilkinton............... "
S. L. Wood .... Dover.
J. Hoyt .... "
L. E. Zearing .... "
Dr. William Robinson ...... "
W. P. Whiteside........East Paw Paw.
J. T. Holbrook ....Arlington.
J. J. Town ....Genesee.
M. S. Brice .... "
N. B. Sweet................ Galva.
L. N. Norton............... Granville.
L. J. Whaler ....Carrolton.
L. E. De Wolf..............Wheaton.
M. Hall.................... "
Henry Bird .... "
H. Smith .... "
D. Kelly .... "
O. J. Reynold ....Mt. Carroll.
A. T. Hyde .... "
A. Tomlinson .... "
James Shaw .... "
E. M. McAffie .... "
B. P. Shirk .... "
O. P. Miles .... "
Adam Bohn................. "
John G. Blake.............. "
David Crouse .... "
V. Armour .... "
A. H. Litchley............. "
H. A. Mills .... "
John Mycum................ "
H. B. Williams........Huntley Station.
A. Hart, Jr. .... Roscoe.
R. J. Cross .... "
L. N Claypool ....Grundy Co.
E. B. Hannah .... "
D. D. Spencer............. "
B. C. Church .... "
J. N. Redding............. "
E. Sanford .... "
J. P. Southerth .... "
Charles Turner........ .... "
C. Grant .... "
G. F. Brown .... "
R. S. Steele............... "
B. Matherson............... "
M. D Hathaway............ Ogle Co.
James Taylor .... "
R. Irwin .... "
M. Hunt.................. "
John P. Taylor .... "
Lewis Payne .... "
E. M. Dutcher............. "
J. N. Edwards............. "

G. N. Henchet .... Ogle Co.
A. H. Treat .... "
J. B. Allen .... "
S. C. Colton .... "
James Rogers .... "
A. N. Swindle .... "
J. V. Gale .... "
A. Barnum ... . "
William Clark.............. "
R. C. Beuchell......... .... "
Levi Harmond............... "
John Conan............... "
William Moore............. "
J. M. Small .... "
H. A. Mix .... "
C. B. Boyce............... "
James Parker............. "
J. N. Mallory .... "
William Fulton .... "
N. P. Bump............... "
F. A. Smith .... "
David Fletcher............. "
E. K. Lister............... "
R. Throop .... "
E. Arnold .... "
S. S. White............... "
S. Ruggles.............. "
Fred. Weyerhauser..... ..Cold Valley.
William Fowler............Middleport.
John D. Donovan .... "
Dr. John S. Pashley........ Oceola.
Capt. Charles Stewart ...... "
R. T. Cassell...............Metamora.
James D. Perry............ "
L. H. Burhans.............. Lincoln.
L. M. Snyder .... "
W. B. Boomer ... . "
George Gates.............Pine Creek.
S. Ruggles .... "
Spooner Ruggles .... "
H. C. Robins..............Kendall Co.
E. Kenbride ....Vermont.
H. S. Thomas .... "
J. H. Hunter............. "
E. C. Gardner............ "
J. Mershom............. "
T. D. S. Stewart .... Reid.
J. F. Dresden .. . . "
H. S. Thomas .... "
Scott Pike................Pittsfield.
N. A. Grimshaw........... "
John S. Irwin .... "
John Webb.... .... "
B. B. Wetz .... "
Stephen Gay.............. "
G. H. Weaver..........Winnebago Co.
William Fisher.............Marshall.
Ira Y. Penn............... "
Richard Loyd .... "

ILLINOIS — *Continued.*

S. S. Richmond............. Marshall.
Ph. Stephens .... "
D. W. Danby .... "
A. Ramsey................. "
F. O. Kenderblake .... "
Joseph Low .... "
John Cook................. "
Lewis Schneider............ "
Henry S. Crane............ "
J. Y. Colton .... "
M. Bangs .... "
John Piper .... "
J. H. Jones................ "
Theodore O. Perry.......... "
N. E. Cook .... "
J. G. Fathershall .... "
J. H. Bond .... "
S. H. Conan................ "
George Scott............... "
D. R. James ....Sycamore.
C. Elwood................. "
H. James .... "
John S. Waterman.......... "
J. M. Hood .... "
J. H. Berridge............. "
C. Kellum .... "
A. Elwood .... "
E. T. Hunt................ "
W. H. Stebbins............. "
M. J. Hunt................. "
H. Safford .... "
J. Moore .... "
N. S. Cathill .... "
Joseph White .... "
H. Elwood................. "
Luther Lowell .... "
Hon. A. L. Knapp, M. C.... Jerseyville.
R. Palmer...... .... Grand Rapids.
Lewis Soule................ "
George Mills .... "
John S. Wilson............. "
M. Wooley................. "
William Ounemeyer......... "
James Packingham .... Granville.
J. F. Alexandre .... "
J. W. Hapgood............. "
Joseph Rheinhart.......... "
Daniel Holly .... "
D. S. Child .... "
S. R. Lewis................Fall River.
J. J. Hanlan .... "
J. V. Hamilton............. Minooka.
H. Kingsly................. Tonica.
A. Spitler .... Carthage.
G. Turner.................. Pekin.
Seth Talbert .... "
S. E. Barber .... "
H. B. Fennington .... "
R. Bergstresser .... Pekin.
George Gregg.............. "
J. D. McIntyre .... "
J. M. Gill.................. "
S. F. Hawley................ "
C. J. D. Rupert............ "
L. H. Case................. "
T. Smith .... "
C. F. Herosey .... "
C. B. Boyce................ "
J. R. Steele................ "
J. M. Mallory............... "
H. Burlingame.............. "
E. C. Stevens.............. Antioch.
F. B. Hopin.............Sangamon Co.
D. C. Hamrow ....Sangamon.
N. Wright.................. "
Abner Edwards ....Bureau Co.
J. T. Holbrook .... "
Smith Miller .... "
John W. White .... "
J. M. Wilson...... .... "
C. L. Collins .... "
Rev. N. S. Collins .... "
John Winter............... "
R. B. Teary............... "
John Crossman............. "
J. P. Knights .... "
W. W. Clark............... "
H. W. Terry .... "
Peter Terry .... "
Henry J. Miller..... .. .... "
Henry Snyder.............. "
J. G. Swann .... "
Tracy Reeves............... "
Julius Benedict............ "
A. R. Kendall.............. "
William Lewis .... "
A. D. Gardner.............. Onarga.
Rev. J. Thomas............. "
S. P. Avery................ "
R. A. Hungerford........... "
Dr. J. L. Parmalee.......... "
Dr. D. Fenner............... "
William P. Pierson .... "
Dr. Silas Earle .... "
R. W. Burt .... Atlanta.
V. C. Secord............... Galva.
C. M. Clark................ "
L. M. Yocum............... "
S. E. Smith................ "
L. N. Miles .... "
F. J. Curtis .... "
James Gallup ....Marshall Co.
Henry F. Ufford ....Lewiston.
J. F. Elrod .... "
William Hall ....Mechanicsburg.
Samuel F. Bowton.......... Sublette.

ILLINOIS — *Continued.*

J. T. Atkinson .... Union Grove.
H. B. Cone .... "
James Pearson .... "
S. W. Bowman............. "
Moses Bross................ "
John Root................. "
J. F. McKebbon .... Freeport.
Francis E. Dakin .... "
J. H. Adams............... "
Henry Story .... "
Levi Dodson ....Champaign.
L. Hodges .... "
A. J. Stone................ "
William S. Waite........... Vandalia.
F. B. Haller................ "
N. M. McCurdy .... "
Andrew Ashton .... Durand.
D. J. Stewart .... "
D. H. Smith................ "
Isaac Sackett .... "
John C. Carman............ Franklin.
H. H. Massey............. Blue Island.
R. Fox .... "
Ernest Ulich............... "
W. P. Roche............... "
H. Massey .... "
H. S. Rexford.............. "
Daniel O. Robinson......... "
Michael Brand .... "
F. Sauterteny .... "
A. Guild .... "
D. J. Pinckney.......... Mount Morris.
John W. Hitt .... "
J. D. Kennedy .... Oswego.
A. P. Grant................ "
B. F. Greene............... "
E. N. Rathbone............. "
A. W. Richardson......... Farmington.
Samuel Cushing........ Creek, Will Co.
John O. Farrell............ Bridgeport.
J. J. Lake............. Clinton Station.
A. W. Martin............... Knoxville.
B. K. Otis................. "
James Cole .... Bushnell.
S. C. Benton............... "
John D. Hall............... "
W. Schooves............... "
U. M. Boor .... "
George M. Epherson......... "
J. Radenbauch .... "
Samuel Eaton.............. Tiskilwa.
A. H. Danby............... "
N. Brilon .... "
R. C. Burhell.............. "
Julius Rogers.............. Warren.
A. L. Brink............... "
E. Clark.................. "
N. Boothly................ "
H. Woodworth .... Danbury.
O. H. Parke................ Macomb.
James McCrosky............ "
William A. Frasier.......... "
Benjamin Smith .... Batavia.
James S. Harvey .... "
T. C. Moore............... "
S. C. Colton............ Grand de Tour.
Henry C. Merrel............ "
C. H. Meland.............. "
William Moore .... Oregon.
Lewis Harmel.............. "
N. N. Jones........... ..... Towanda.
D. L. Zebraskia............ St. Charles.
J. P. Tumarld .... "
E. S. Holbrook............. "
John M. Cabliegh........... "
H. House.................. "
Read Feison .... "
L. D. Morgan .... "
S. S. Jones .... "
Thomas Cooth.............. Dunleith.
T. H. Brittan............... Verdan.
C. L. Eaton... .... Homer.
W. H. Tupper.............. Kaneville.
Lewis Stewart.............. Plano.
H. Bronson................ Princeville.
J. B. Merritt................ Rockton.
A. N. Gilman .... "
S. J. Goodwin .... "
J. C. Crumb .... Harvard.
A. E. Axtell............... "
L. B. Wyant .... "
T. D. Pierce .... Addison.
T. J. Evans .... Cobden.
R. Freeman................. Athens.
H. Coleman................ Pine Rock.
H. N. Owen................. McHenry.
J. H. Straight.............. Fairboro.
R. C. Straight............. "
L. E. Crittenden............ "
John Scott................. Tremont.
Edwin Littlefield .... "
J. B Wait................ Brookfield.
W. Holton............. Willow Creek.
R. T. Butler .... Viola.
Nathan Oug .... Magnolia.
Robert Tweed.............. Ringwood.
T. M. Kimball............. "
H. R. Keightly .... Knoxville.
O. H. Ellethorp .... Burlington.
S. H. Whaples .... Neponset.
Moses Cherry .... Oswego.
M. J. Cummings............ "
Willis Nelson .... "
J. J. Coles................ "
C. S. Roberts .... "
C. L. Roberts .... "
T. Coffin.................. "
Albert Snook.............. "

ILLINOIS — *Continued.*

Name	Place
William Cowdrey	Oswego.
H. L. Hudson	"
Thomas D. Wayne	"
H. C. Hopkins	"
William L. Fowler	"
Henry McKenny	"
William B. Bloomer	"
S. J. Minkler	"
E. S. Hobart	Tazewell Co.
E. Dillon	"
H. R. Greene	"
James M. Thomas	Stark Co.
J. Thomas	"
Benjamin Turner	"
William P. Pierson	Iroquois Co.
James De Wolfe	Carroll.
E. Worthy	"
J. Jones	"
S. L. Choate	Buda.
M. F. Dewitt	Greene.
J. B. Samuel	"
Henry Hosmer	Chilicothe.
S. J. Gilbert	"
P. T. Mathews	"
William M. Davenport	"
O. W. Young	"
William McClure	"
M. Dupies	Savanna.
A. H. Treat	Foreston.
J. B. Allen	"
W. H. Robbins	"
George W. Hewett	"
Peter R. Reid	Rock Island Co.
John Dewe	"
John M. Gould	"
J. T. Richards	"
G. B. Shaw	"
Hiram Pitts	"
L. C. Boynton	"
Charles Atkinson	"
John Dahmes	"
Truman Humphries	"
J. H. Pickerell	Macon Co.
John G. Taylor	"
L. Burrows	"
E. W. Smith	"
A. A. Hardy	"
Henry Prather	"
D. S. Allen	"
William Howard	Scott Co.
Robert C. Beech	"
D. D. Brugle	"
G. G. Terry	"
T. F. Cooke	"
S. P. Coons	"
J. P. Hayes	Putnam Co.
D. S. Childs	"
J. Packenham	"

Name	Place
H. W. White	Putnam Co.
A. P. Drysant	"
C H. Whitaker	"
W. H. Brown	"
William Eddy	"
N. Sherwood	"
W. A. Punnell	"
W. T. Cuninngham	Vermillion Co.
J. Peters	"
S. M. Sullivan	"
R. V. Chesley	"
M. D. Hawes	"
C. K. Myers	"
A. N. Chadwick	Franklin Grove.
G. M. Brayton	"
J. W. Edmonds	Nachusa.
W. A. Rulley	"
Benjamin Turner	Stark Co.
William Lowman	"
J. M. Thomas	"
Dr. C. Gorham	York.
John W. Sewell	De Kalb Co.
J. F. Glidden	"
Z. Nobles	"
E. H. Smith	"
R. Hopkins	"
Leonard Moore	"
J. F. Simonds	"
H. R. Francisco	"
F. C. Hapgood	"
J. P. Earl	"
D. E. Adams	Lena.
Col. P. Stewart	Wilmington.
C. B. Stetson	"
David U. Cobb	"
Franklin Mitchell	"
Hiram O. Alden	"
Edward Alden	"
Rev. C H. Force	South Ottawa.
S. Crook	"
J. H. Dickens	"
William Brown	"
M. D. Calkins	"
Evander Fields	"
H. Mayo	"
John Hill	Petersburg.
J. W. Scroggs	Champaign City.
C. H. Miller	"
John Dahmes	Moline.
Hiram Pitts	"
A. P. Dysert	Hennepin.
John B. Fairbanks	Morgan Co.
James R. Hammet	Tuscola.
H. T. Carnaway	"
G. F. Ward	Milton.
George Wells	Lake Co.
George D. Williston	Calumet.
W. C. Flagg	Alton.

WISCONSIN.

J. D. Culver .... Wisconsin.
L. D. Blossom............. "
James Whitney .... "
G. H. Paul .... "
C. S. Kelsey............... "
L. G. Fisher .... "
B. J. Stevens .... "
J. A. Davis .... "
D. Bonestead .... "
S. Clark .... "
George Wickett .... "
K. H. Darling .... "
John Peacock .... "
S. S. Sanborn.............. "
Capt. J. Tuttle .... "
P. Judson .... "
H. S. Thorp................ "
S. B. Husselman .... Madison.
James E. Morsley .... "
O. Cole .... "
C. L. Williams .... "
J. E. Bird .... "
E. S. Carr .... "
E. M. Keys .... "
R. J. Stevens .... "
A. J. Darin .... "
Lewis B. Viles .... "
W. B. Van Slyke .... "
F. D. Fulton .... "
James Ross................ "
James Richardson .... "
Daniel Real................ "
George P. Daplin .... "
John V. Robbins............ "
F. Briggs.................. "
L. F. Brownell .... "
Alfred Merrill.............. "
Charles M. Cushman........ Milwaukee.
D. McDonald............... "
E. Kahn.................. "
Matthew Watkins .... "
A. Mullen....... .... "
James B. Kellogg .... "
J. A. Helfenstein .... "
A. C. May .... "
J. Hochstrader............ "
J. A. Noonan .... "
John Nazro .... "
J. J. Benehan .... "
H. Bartlett................ "
C. H. McKeever............ "
H. L. Palmer............... "
Samuel J. Hooker .... "
Ed. H. Broadhead .... "
Charles Gasbury............ "
John Thompson .... "
R. P. Sanders.............. "
John Rice .... "
J. H. Cordes .... "
Charles Leesburg.......... Milwaukee.
E. M. Randell............. "
Gen. John McManman ...... "
J. Graham .... "
G. H. Walker.............. "
E. L. Jones .... "
H. B. Merrill .... "
R. G. Clarkman .... "
S. P. Tilton................ "
S. Hyatt ... . "
F. A. Platt .... "
Samuel Dall, Jr............. "
Anson Ballard .... Appleton.
Peter White .... "
T. M. Woodward .... "
George McDonald .... "
Alvin Foster .... "
W. S. Warner............... "
Francis Hammond .... "
Prof. R. Z. Mason .... "
John S. Lester.............. "
William Johnson........... "
E. P. Humphrey............ "
Henry D. Ryan............. "
George M. Robinson ... . "
Moses M. Davis .... "
A. L. Smith................ "
T. R. Hood .... "
E. C. Gough .... "
Byron Douglas .... "
J. H. Kelly .... Racine.
L. J. Blake .... "
P. Thorp .... "
N. J. Van Pelt............. "
S. E. Hurlburt............. "
J. J. Case .... "
A. P. Dutton .... "
John Tapley .... "
D. Andrews................ "
W. Fitch.. .... "
George C. Northup.......... "
T. J. Emerson............. "
H. G. Winslow............ "
S. C. Tuckerman............ "
J. C. Paine .... "
W. T. Van Pelt .... "
C. H. Upham .... "
A. Sanford .... "
D. H. Jones ... . "
George A. Thomson .... "
T. Tayler .... "
C. A. Lathrop.............. "
T. Talvey.................. "
H. J. Ulham .... "
W. W. Vaughan............ "
M. Kiedal................. "
M. B. Mead............... "
A. J. Wickham .... "
J. G. Meacham .... "

WISCONSIN — *Continued.*

Name	Place
John Dickenson	Racine.
R. H. Mills	Beloit.
John G. Fisher	"
W. A. T. Chapman	"
J. J. Blaisdall	"
G. W. Bicknell	"
W. C. Ritchie	"
B. E. Hall	"
G. More	"
J. C. Newcomb	"
G. H. Stocking	"
S. T. Murrell	"
B. C. Rogers	"
J. Manchester	"
John Hackett	"
S. W. Peck	"
J. H. French	"
J. G. Kendall	"
E. D. Murry	"
H. Rosenbault	"
E. Abel	"
Joseph Brittain	"
J. G. Winslow	"
L. G. Fisher	"
O. Manchester	"
Volney French	Kenosha.
F. Robinson	"
H. F. Schoff	"
O. T. Head	"
Z. G. Simmons	"
A. Farr	"
H. H. Tarbell	"
F. N. Lyman	"
H. Durkee	"
C. Durkee	"
A. B. Smith	"
M. Frank	"
E. Bain	"
F. H. Head	"
William E. Read	"
E. H. Kellogg	"
G. T. Van Osdale	"
Charles C. Sholes	"
D. Stone	"
John Tuttle	"
Edward H. Rudd	"
P. Judson	"
G. F. Mosley	Janesville.
W. A. Laurence	"
C. Miner	"
S. D. Burlingame	"
A. P. Prichard	"
C. C. Keeler	"
William Vankirk	"
J. H. Vankirk	"
John N. Lynch	"
R. B. Trust	"
E. D. Murdock	"
T. R. Carswell	"

Name	Place
P. H. Brady	Whitewater.
L. A. Tanner	"
R. O'Connor	"
Alexander Graham	"
H. L. Paher	"
L. A. Winchester	"
George Esterly	"
A. B. Jackson	Menasha.
A. L. Collins	"
J. G. Calkins	"
E. D. Smith	"
Dr. Whittlesy	"
M. Hoffan	"
M. Selles	"
G. R. Perry	"
Fred S. Ellis	Green Bay.
Daniel M. Whitney	"
Otto Frank	"
C. D. Robinson	"
H. G. Turner	"
J. S. Doge	Darien.
N. H. Hoag	"
T. Phelps	"
W. H. Whitney	Geneva.
W. B. Walker	"
H. Allen	"
C. Miller	"
C. A. Stephens	La Crosse.
Charles W. Marshall	"
Charles Seymour	"
C. C. Moore	Mineral Point.
Moses M. Strong	"
L. D. Lake	Green Lake.
C. J. Oatman	"
F. C. Smith	"
S. H. Stafford	"
C. M. Baker	"
T. D. Hale	"
Frank Higgins	Columbus.
James Maxwell	Baraboo.
A. W. Stark	"
A. Weatherby	Shullsburg.
B. Blake	Port Washington.
M. Fellows	Manitowoc.
Joseph J. Vilas	"
James Edwards	Fond du Lac.
B. J. Munn	"
James Jenks	Oshkosh.
S. M. Hay	"
M. E. Smith	Fox Lake.
E. P. Esta	Elkhorn.
S. H. Ashum	Waupaca.
A. J. Lyman	Sheboygan.
M. R. Young	Grant Co.
J. M. Bingham	Palmyra.
James G. Thorp	Eau Clair.
H. L. Danton	Prairie du Chien.
J. H. Warren	Albany.
George A. Dill	Prescott.

WISCONSIN — *Continued.*

C. Hoeflinger .... Warsaw.
N. H. Wood............. Portage City.
J. F. Myer............... Watertown.
E. C. Lews .... Juneau.
H. B. Hawley .... Jefferson.

IOWA.

John Hume.... .... Davenport.
A. Struss .... "
E. S. Gilbert .... "
James E. Henry .... "
E. Cook .... "
L. Shriecker .... "
George L. Davenport........ "
Richard B. Hill .... "
David Higgins.............. "
E. Mack.................. "
W. F. Washburn .... "
Gen. Fitz Henry Warren.... Burlington.
John H. Geere .... "
E. E. Gay................. "
O. H. Schank .... "
E. H. Thomas .... "
F. W. Brooks .... "
L. S. Ellithorp .... "
L. H. Shephard .... "
Samuel Rand............... Lyons.
J. P. Gage .... "
S. J. Magill .... "
H. H. Harrison .... "
Henry G. Hill .... "
N. Boardman............... "
Thomas Crew.............. "
James Birney .... "
J. B. Miller................ "
Charles Ryan .... "
David Joice................ "
Hiram Gates............... "
J. C. Wann............... "
Charles Brayton............ "
G. A. Buffam .... "
George H. Jerome.......... Iowa City.
E. Shephard .... "
H. D. McKay .... "
P. Musser .... "
M. T. Durant .... "
G. E. De Forest .... "
James H. Gower .... "
S. E. Paine .... "
C. T. Rawson .... "
E. M. Chase................ "
David W. Kilbourne .... Keokuk.
Edward Kilbourne.......... "
William Leighton .... "
George Williams, Jr......... "
D. L. McGrugan............ "
William Timberman .... "
M. J. Stannan.............. McGregor.
W. J. Gilchrist............. "
J. N. Vanoman............ "
A. P. Richardson .... McGregor.
R. Stout .... "
J. Bassett .... "
A. Waggener .... "
J. Brown .... "
D. D. Hosix................ "
John McGregor .... "
E. R. Brown .... "
J. F. Bassett .... "
George W. Bradenburg...... Ottumwa
George Murry.............. "
J. M. Morris........... .... "
James Hawley.............. "
J. A. Drake................ "
Dr. W. P. Corran........... "
J. W. Edgeley.............. "
J. Chambers .... "
W. D. Wilson............ Des Moines.
L. W. Dennis .... "
F. Mills.................... "
L. H. Bush .... "
George Whitaker .... "
J. M. Needham .... Oskaloosa.
J. Thompson.............. Cedar Falls.
J. Vendegraff.............. De Witt.
W. Fuller .... "
J. F. Homer................ "
B. F. Page .... "
Joseph Howe............ Mt. Pleasant.
S. Herrick .... "
Dr. J. H. Fulton .... "
W. A. Sanders .... "
W. P. Sanders .... "
Warren Blue .... Lansing.
M. Van Winter .... "
F. C. Bowen............ Independence.
P. C. Wilcox............... "
R. R. Plance .... "
John F. Cook .... "
D. S. Durham .... "
G. M. Woodbury......... Marshaltown.
T. Babel .... "
George Glick .... "
E. W. Lockwood .... "
H. P. Williams............. "
A. C. Camp .... Cedar Rapids.
F. M. Brown................ "
A. A. Lawn................ Irving.
C. Schwaller............... Clayton.
H. H. Packard .... Decora.
W. C. Woodworth .... Fort Madison.
J. J. White............... Winterset.
John Waggener............ Burr Oak.

IOWA — *Continued.*

B. C. Hopkins.........Bowen's Prairie.
Otis Whitmore............. "
A. H. Fox................West Union.
Lewis Berkley............. "
S. Carpenter..............Bloomfield.
A. Hart.................. "
R. Johnson.............Princeton.
J. B. Grinnell.............Grinnell.
Marshal Bliss.............. "
James Motheral............Macomb.
Dr. J. Carey..............Toledo.
E. P. White..............Muscatine.
G. L. Bell................ "
J. M. Dally...............Magnolia.

MINNESOTA.

John W Taylor............St. Paul.
J. F. Wierst............. "
R. A. Mott................Faribault.
James H. Parker............Red Wing.

MISSOURI.

Hon. J. C. Filly............St. Louis.
J. Cheever............... "
C. W. Horne.............. "
A. W. Fredin............. "
Benjamin Charles.......... "
G. W. Dyer............... "
Henry T. Blow............ "
C. D. Drake.............. "
B. R. Bonner............. "
L. B. Hayden............. "
C. W. Irwin.............. "
Gen. Wm. K. Strong........ "
A. J. Brown.............. "
G. W. Curtis............. "
A. B. Morien............. "
E. W. Fox................ "
E. O. Steward............ "
Wm. M. McPherson......... "
L. D. Hayden............. "
S. H. Laflin............. "
Wyman Crow............... "
Charles M. Irvine......... "
B. G. Brown.............. "
George Patridge........... "
A. G. Brown.............. "
Alexander B. Marceau...... "
Daniel R. Garrison........ "
Ed. Wider................ "
J. J. Homer.............. "
S. M. Edsell................St. Louis.
H. A. Hosmer............. "
C. S. Grady.............. "
James Richardson.......... "
George P. Strong.......... "
W. K. Strong............. "
Alfred Pierce............. "
T. H. Homer.............. "
S. A. Homeyer............ "
C. S. Greely.............. "
Morris Collins............ "
Henry Hitchcock........... "
B. Gratsbrown............. "
William S. Mosley......... "
S. H. Boyd............... "
Washington Adams.......... "
Phillip Drake............. "
George H. Nettleton........Hannibal.
Henry Staring............. "
James Craig................St. Joseph.
W. S. Osley...........Jefferson City.
M. S. Mosley.............. "
J. H. Boyd................Springfield.
W. Adams..................Boonville.
W. J. Croch...............Livingston.
D. W. C. Edgerton......... "
B. P. Henno..............Chillicothe.
Dr. J. B. Bed............. "

KENTUCKY.

N. J. Norton..............Shelby.
J. B. Knight.............. "
W. J. Knight.............. "
Theodore S. Page.......... "
A. J. Bradley..............Scott Co.
Dr. A. Rodman.............Frankfort.
R. H. Crump...............Louisville.
Wm. Terry................. "

KANSAS.

J. Rice.................Leavenworth.
R. J. Hirton.............. "
A. M. Sawyer.............. "
Dr. Lee Huston............ "
Hon. Mark Delahny......... "
Hon. M. J. Parrott.......Leavenworth.
Gen. John A. Halderman.... "
Col. Chas R. Jennison..... "
D. W. Wilder.............. "
Ed. Conservetive.......... "

KANSAS — *Continued.*

J. L. Pendry	Leavenworth.	C. F. Currier	Leavenworth.
D. H. Baily	"	J. Ingersol	"
H. Buckingham	"	Dr. J. Jenks	"
J. Sands	"	Dr. C. A. Logan	"
J. H. Moore	"	Dr. S. K. Henderson	Lawrence.
Ed. Woodruff	"	J. C. Trask	"
J. C. Irwine	"	Dr. D. W. Stornet	Topeka.

DISTRICT OF COLUMBIA.

J. L. McKee	Washington.	Wm. A. McCan	Washington.
E. H. Heal	"	Charles A. Page	"
Peter Parker	"	L. D. Personate	"
J. R. Briggs	"	James E. McClean	"
F. A. Springer	"	C. F. Cook	"
Ward H. Lamon	"		

DACOTAH TERRITORY.

William Jayne.

[The following letters were mislaid, but having been recovered are here appended:]

FROM ADMIRAL JOHN A. DAHLGREN,

Commander of the Naval Forces before Charleston.

BUREAU OF ORDNANCE, NAVY DEPARTMENT,
WASHINGTON CITY, May 9, 1863.

Hon. ISAAC N. ARNOLD, Chicago:

MY DEAR SIR:—Your kind note of the 1st has just reached me.

The object which your Convention has in view is truly important, and I judge from the names of those who sign the call, that it is likely to receive the attention it deserves.

I regret that the demands on my time by public duties will most probably prevent me having the pleasure of being present at the meeting.

In the course of business, I have just visited some places on the Ohio and Mississippi where vessels are being built for service on these rivers. It was the first time I had been west of the mountains, but in the brief duration of a hurried journey, I was able to realize the impression I had formed of the great future of the West.

Facility of commerce for products of all kinds is absolutely essential to the due development of its resources, and to this, the object you have in view is one that must conduce very materially.

I wish much that my views were sufficiently matured to give them expression in writing, but the subject embraces too much that is wholly unknown to me, and which would be indispensable to any statement that could be of the least serivce.

With my best wishes for the success of this and every other project that may add to the prosperity and union of our glorious country, I am most truly,

Your obedient serv't,

JOHN A. DAHLGREN.

FROM THE LATE ADMIRAL A. H. FOOTE,

Of the United States Navy.

WASHINGTON, D. C., May 24, 1863.

MY DEAR SIR:—Your kind letter of invitation of the 1st instant, with its enclosed call for a national convention, has been received. With the pressure of my public business I am unable to leave Washington for the present, unless it be on a tour of inspection at the different navy yards and the rendezvous, otherwise I should be most happy to avail myself of your kindness; not that I consider myself at all competent to render aid in promoting the laudable object of your Convention, but rather to see Chicago and my many friends in your prosperous city.

While I take a deep interest in everything tending to promote our national prosperity by the development of its commercial facilities, and strengthening the bonds of union, I trust that the objects of your grand Convention may be consummated. The character of my pursuits in life, and limited information on the subjects which are to come before you, hardly warrant me in preparing a paper to be read before the Convention, with the idea that it could prove of any value.

Thanking you for the compliment, and with my respects and kind regards for Mrs. Arnold,

I have the honor to be, very respectfully, your obedient servant and friend,

A. H. FOOTE.

Hon. I. N. Arnold, &c.

FROM HON. WM. SPRAGUE,

A United States Senator from Rhode Island.

Providence. R. I., May 1, 1863.

Dear Sir:—I have your kind note of the 1st. I will try and be with you. I desire to be second to none in interest for your enterprise. Truly yours,

Hon. I. N. Arnold, Chicago. WM. SPRAGUE.

FROM HON. A. B. OLIN,

Chairman of the Military Committee of the House of Representatives of the 37th Congress.

Washington, D. C., May 4, 1863.

Hon. I. N. Arnold:

My Dear Sir:—I have received your note and the accompanying circular of the proposed commercial Convention, to be held on the first Tuesday of June next.

I intend, with God's blessing, to be there. I feel a very deep interest in it. I regard my life and labors in Congress as a substantial failure by reason of the loss of our bill.

I am sure that if that act had passed, it alone would have given the last Congress a better title to the respect of posterity than any or all of its other acts. I have no doubt the proposed Convention will so prepare the public mind, as to make your triumph certain and your victory easy at the next session of Congress. I feel no regret at leaving Congress, and yet I confess I should like much to share with you the distinguished honor which will attend the success of a measure of such transcendent importance to the country, as I regard this to be.

Very truly yours,

A. B. OLIN.

FROM HON. SCHUYLER COLFAX,

A Member of the House of Representatives from Indiana.

WASHINGTON, D. C., May 12, 1863.

GENTLEMEN:—If the obligation of duty and affection toward a suffering member of my family here, would allow me to obey the call of duty and of patriotism which beckons me toward your great Canal Convention at Chicago next week, I should be there in person instead of by this epistolatory substitute. None of the other conventions of this year, except those which assemble together the friends of the Union for united action, rivals this one in its far reaching and vital importance; and as the Mississippi flowing day and night and month and year from north to south, unceasingly protests against the separation of these sections from each other, so the great enlarged canals, creating another Mississippi from west to east, will strengthen the bond between the seaboard and the frontier, and give to that fervid patriotism which declares that the whole Union must remain intact forever, the potential aids of commercial policy and industrial interest.

The nation also will reap a rich pecuniary reward from the completion of this great enterprise; but I can glance hurriedly at but one or two points. When the shipment of produce, by the sale of which the farmer lives, from the frontier to the ocean, costs one-third to one-half, and even three-fourths of the whole amount realized for it when it reaches the wharf, the producers must necessarily be poor customers for all these goods by the duties on which our treasury is in a large degree replenished; but with the cost of transit reduced, as it will be by the broad artificial river to the sea which you contemplate, the prosperity among our agriculturists which will follow, will be felt in that result on the treasury receipts, which can be predicted with such unerring certainty again.

When the United States Tax Assessor computes the invoice of the farmer, and values the corn at ten or twenty cents per bushel, he can find but few that will be required to aid the treasury tax on their surplus profits.

But with the West, by decreased cost of transportation, brought in prices so much nearer the East, the increased taxes on the increased wealth resulting from your improvement, will indicate its wisdom even as a pecuniary investment.

Confident that your assemblage will conduce by the unity and wisdom of its deliberations and action, to the speedy accomplishment of a work that cannot fail to promote the development, prosperity and unity of our whole country,

I am respectfully yours,

SCHUYLER COLFAX.

Committee of Invitation of the Chicago Convention.

MEETING

OF THE

NATIONAL CANAL COMMITTEE.

Pursuant to notice, the Executive Committee appointed by the President of the Convention, met at the *St. Nicholas Hotel*, New York City, on Wednesday, October 7th, 1863, at 3 o'clock, P. M.

Mr. I. N. ARNOLD, of Illinois, Chairman, called the Committee to order, and in the absence of Col. FOSTER, R. B. HILL, Esq., of Iowa, was elected Secretary.

The States of Illinois, New York, Ohio, California, New Jersey, Vermont, Massachusetts, Indiana, Minnesota, Kansas, Maine, Wisconsin, Kentucky, and Rhode Island, were represented.

The Sub-Committee, to prepare a Memorial to the President and Congress of the United States, submitted a draft of a Memorial for the consideration of the Committee.

The Memorial was read to the Committee, and after being amended, on motion was unanimously adopted.

On motion of Mr. HILL, it was

Resolved, That a Committee of five, consisting of the Chairman, Mr. ARNOLD, and four others to be named by him, be appointed to present the Memorial to the President of the United States, and ask him to lay the same before Congress, with a recommendation that Congress adopt the most efficient means to secure, as early as practicable, a ship-canal from the Mississippi to Lake Michigan, and from the Lakes to the Atlantic.

The Chairman named as members of the committee, JUSTIN S. MORRILL, of Vermont, JAMES A. McDOUGALL, of California, A. A. LOW, of New York, and RICHARD B. HILL, of Iowa.

Whereupon the Committee adjourned, to meet at the call of the Chairman.

ISAAC N. ARNOLD, *Chairman.*

RICHARD B. HILL, *Secretary.*

MEMORIAL

TO THE

PRESIDENT AND CONGRESS OF THE UNITED STATES.

The assemblage of the National Canal Convention, so great a gathering of the people, in the midst of such a war as that which is now taxing the resources and energies of the American people to the utmost, was in itself a striking and significant fact.

This Convention, national in its objects and its numbers, connects itself in the minds of all thoughtful men with the political unity of the country. An instinctive conviction of its great importance and direct bearing on the national unity, secured for the call for the Convention a hearty and cordial response, scarcely paralleled in the past history of our country.

The meeting also indicated the conviction in the minds of the whole people, of the existence of a great *need*, profoundly realized, and a determination to supply that necessity. That need is, *enlarged water-facilities for communication between the East and the West*, both for military and commercial purposes. The subject of enlarging canals between the valley of the Mississippi and the Atlantic was evidently regarded as the great question of the times, excepting always the duty of putting down the rebellion, and maintaining our national integrity.

Under the resolutions of the Convention by which this Committee was raised, our duty, as we conceive, is, not to designate the manner in which ship and steamboat channel or channels may be opened between the Mississippi and the Atlantic, but to present the views of the Convention upon the general subject to the President and Congress, leaving it for the Government itself, in its wisdom, to determine the best and most judicious plan of effecting the great object.

The Convention was entirely unanimous in the resolution, that the construction or enlargement of the canals between the Mississippi and the Atlantic, with canals connecting the Lakes, was of great national, military, and commercial importance, and that such enlargement to dimensions adequate to pass gun-boats from the Mississippi to Lake Michigan, and from the Atlantic to and from the Great Lakes, would furnish the cheapest and most efficient means of protecting the Northern frontier, and at the same time would promote the rapid development and permanent union of our whole country.

Your memorialists in presenting the views of the Convention to the Executive and Congress, will not attempt to go into details; they refer to the mass of facts and statements contained in the able reports of the Boards of Trade, and in the letters, surveys, etc., presented to the Convention and embodied in its published proceedings.

NECESSITY OF SHIP-CANALS BETWEEN EAST AND WEST.

The one great idea which your memorialists seek to impress upon Congress is, the necessity of a great national highway, in the form of a ship and steamboat-canal between the Mississippi and the Atlantic.

This great *national highway* is demanded alike "by military prudence, the necessities of commerce, and political wisdom."

Your memorialists ask the attention of the Government to some of the reasons, *military*, *commercial*, and *political*, why this work should be constructed.

1. The Military Importance of the Work.

We have arrived at that period in our history, in which the Government should adopt a well-considered and systematic plan of defending the Northern frontier. Indeed, the best means of doing this has long engaged the attention of the Government. Reports from the War Department, and surveys from the Topographical Corps, in great numbers, have been made, a large number of forts have been projected and surveyed, but little has as yet been done.

The importance of having command of the Lakes, in case of a war with Great Britain, cannot be over-estimated. In 1814, the Duke of Wellington declared that "a naval superiority on the Lakes is a *sine qua non* of success in war on the frontier of Canada." The great military importance of the command of the Lakes was illustrated in the war of 1812, and the victories of Perry on Lake Erie, and of McDonough on Lake Champlain, were decisive of the fate of the war on the northern border.

In our past history, in the old Colonial and Revolutionary wars, and in the late war with Great Britain, the principal attacks our country had to sustain were made from the Canadian frontier. But the defense of the Northern frontier, always of great moment, has become, by the growth of the West, of incalculable importance. Certainly not less than one-third in value of the entire commerce of the nation passes over the Lakes. Ten millions of people live upon their borders, and are directly interested in their security. The great cities, which have grown up on their shores, have become the largest grain depots of the world. Nowhere on earth are collected and distributed such vast amounts of food; and yet this commerce, vast as it is, these great cities and food-producing States, with their great granaries, lie entirely exposed, and invite, by their helpless condition, ravage and devastation.

We say confidently, that this condition of things will not be permitted to continue. The voice of the North-West and of all the Northern frontier will ask, (and their just request will be cheerfully granted,) adequate protection.

EXPOSED CONDITION OF NORTHERN FRONTIER AS COMPARED WITH THE ATLANTIC COAST.

We respectfully call the attention of Congress to the defenseless condition of the Northern frontier, as compared, or rather as contrasted, with that of the Atlantic. Upon the defenses of the Atlantic, exclusive of naval defenses, there have been expended more than one hundred and fifty millions of dollars. Large additional appropriations were asked for and obtained at the last session of Congress, and yet the Atlantic shore is more than three thousand miles from a foreign foe. An ocean shields it from attack. It is defended by the strongest navy possessed by any nation on earth. For all this we pay cheerfully, nor do we question the propriety of these expenditures; we only ask that the Northern line shall be no longer neglected.

Now, we earnestly call attention to the fact that our Northern frontier, with its commerce, and cities, equal in value to the seaboard, and with a shore-line exceeding in length the Atlantic coast, is within rifle and cannon range for a considerable distance, of the only great maratime nation which will ever give us serious trouble, and is entirely defenseless. We have no navy on the Lakes, nor can we have under existing treaties. We have neither forts nor fortifications, nor ordnance, nor navy yards. Our Northern frontier is utterly without the means of defense.

DEFENSES OF CANADA.

Let us contrast our means of defense with those of our neighbors over the line. In 1817 it was provided by treaty between

Great Britain and the United States, that both nations should dismantle their vessels of war on the Lakes, and reduce their naval force on each side "to one vessel of one hundred tons burden on Lake Ontario, and one on Lake Champlain, each armed with one eighteen pound cannon, and on the upper Lakes to two such vessels armed with the like force."

Since this treaty, Great Britain has never lost sight of the security of her American Colonial Empire. She has expended many millions for its defense. She has large and important military defensive works at Kingston on Lake Ontario, at Malden at the mouth of the Detroit river, at Penetanguishene on Georgian Bay, at Toronto, Niagara, Stanley, Windsor and Port Sarnia, and others extending west as far as the shores of Lake Superior, and beyond to Fort Williams and Fort Gary.

GREAT BRITAIN RELIES ON HER MILITARY CANALS, CONSTRUCTED FOR MILITARY PURPOSES.

But the main reliance of England for maintaining and securing her supremacy on the Lakes, is upon her military canals. These she has constructed at great expense, to enable her to pass her gun-boats from the ocean through the St. Lawrence to the Lakes. These works were constructed with direct reference to their military uses.

The canals from Montreal, by way of the Ottawa river and interior Lakes, to Kingston on Lake Ontario, were constructed avowedly as a military work by the Royal Engineers, under the direction of the British Ordnance Department. The preamble of the act of the Canadian Parliament authorizing the taking of lands for the purpose, recites, that

> "His majesty has been pleased to direct measures to be immediately taken, under the superintendence of the proper *military department*, for constructing a canal connecting the waters of Lake Ontario with the Ottawa river, and affording a convenient navigation for the transport of *naval* and *military stores*."

In 1831, Col. Dumford, of the Royal Engineers, in his testimony before a committee of the English Parliament, stated, that provision had been made for the defense of the Lakes, and the canal being intended as a *military* work, fortifications should be erected at the entrance of the canal, and the immediate vicinity at Kingston. A fortress of very considerable strength has been built at Kingston. This canal was followed by the construction around the rapids of the St. Lawrence of a series of short canals, far transcending in capacity any commercial necessity at the time they were built, with locks 45 feet wide, by 200 feet long, and 8 feet deep.

She has also connected Lakes Erie and Ontario by the Welland canal, of great capacity, with locks 26 feet wide, 150 feet long, and water 11 feet deep.

Such are the means by which Great Britain, sagacious and persistent, and ever looking to the possibility of war, has provided for securing the control and supremacy on the Lakes.

It was the confidence growing out of the condition above described, and a knowledge of our own defenseless condition, that induced the *London Times*, during the excitement growing out of the seizure of Slidell and Mason, to publish articles like the following:

"ARMING THE NORTHERN FRONTIER AND THE LAKES."

"The worst part of the struggle, however, will not be on the Atlantic seaboard, but on the Great Lakes of Upper Canada and North America. We are glad, therefore, to be able to tell our readers that *this danger has been foreseen*, and amply provided against, and that within a week after the breaking of the ice a whole fleet of gun-boats with the most powerful of screw corvettes sent out to Admiral Milne, will carry the protection of the British flag from Montreal to Detroit."

The exposed condition of the Lakes has not escaped the attention of Congress. The Committee on Military Affairs of the House of Representatives, through Hon. F. P. Blair, Chairman, at the last session of Congress, called the attention of the country to this startling statement:

"A small fleet of light draft, heavily armed iron-clad gun-boats could in a short month pass up the St. Lawrence into the Lakes, and shell every city from Ogdensburgh to Chicago."

"It could at one blow sweep our commerce from the entire chain of waters. Such a fleet would have it in its power to inflict a loss to be reckoned only by hundreds of millions of dollars, so vast is the wealth thus exposed to the depredations of a maratime enemy."

The cost of all the improvements proposed in this connection would be but a trifle, compared with the loss which could be inflicted by a single *raid* by these gun-boats through the Lakes. This condition of exposure must not be permitted to continue; and thus we are brought to the question of

WHAT ARE THE BEST MEANS FOR DEFENDING THE LAKES?

Two plans have been proposed; one, that of a chain of forts along the shore, defending the entrance of each lake and other strategic points, and fortifications for the security of each considerable town and city.

The other is the construction or enlargement of such canals as will enable our fleets of gun-boats to pass from the Ocean by the Mississippi and the Hudson to the Lakes. The objections to the former are grave, and in the judgment of your memorialists, viewed in the light of the experience of the present, are decisive.

The expense would be enormous, and when constructed, as against iron-clad gun-boats, would prove unavailing. The war in which our country is now engaged has shown, especially have the seige of Charleston and the comparative vulnerability of Forts Wagner and Sumter demonstrated, that earthworks are better than regular walled forts, and that neither are adequate to prevent the passage of iron-clad gun-boats.

We must, if practicable, do as Great Britain has done—construct military canals, adequate in capacity to admit our gun-boats to the Lakes. Thus we shall be placed upon an equality with our neighbors.

Fortunately, this is entirely practicable, and with but small expense as compared with the important results to be secured.

Various plans for constructing and enlarging canals, to enable gun-boats to pass from the Hudson and the Mississippi to the Lakes, have been suggested.

Prominent among others, is that of enlarging the present Illinois and Michigan canal from Lake Michigan to Lake Joliet, a distance of only thirty-six miles, and the improvement of the Illinois and Desplaines rivers, so that all steamers and gun-boats which navigate the Mississippi, can pass directly into Lake Michigan at Chicago. Also, a ship-canal around the Falls of Niagara, and the enlargement of the locks of the Erie and Oswego canals, by which gun-boats can pass directly from the Atlantic into Lakes Ontario and Erie.

We will not undertake to decide between the merits of these various propositions, nor whether it may not be expedient to enter upon the construction of all, or to extend aid to all; but we would most earnestly press upon the consideration of the President and of Congress, the importance of securing at the earliest practicable period, a steamboat and ship-canal from the Mississippi to the Lakes, and from the Lakes to the Hudson and the Atlantic.

It was well said by Washington, "that if we desire peace, it must be known that we are at all times *ready for war.*"

The military position is, in a few words, this: On the American side, the Northern frontier is defenseless. It is amply defended on the British side. England can take her gun-boats from the ocean through the canals and the St. Lawrence into the Lakes with facility. We cannot do it at all. Great Britain has constructed canals for this express purpose. We have no such military canals. England, years ago, did that for the defense of Canada, which our Government is now asked to do for our own country. Without the ship and steamboat-canals, our Lake commerce and cities are at her mercy. With the enlarged canals, through our great superiority in mariners, steamers, vessels, and material on the Lakes, we are secure, and our supremacy is certain.

Our security will be found in providing the means of floating "*Uncle Sam's webbed feet*," as the President calls our gun-boats, into the Lakes. The work done by these "web-footed gun-boats" in this war, "not only on the deep sea, the broad bay, and the rapid river, but also up the narrow bayou and wherever the ground was '*a little damp*,'" has furnished most valuable illustrations of their importance in all military operations.

Surely our Government will not do less in providing military canals for the security of the very heart and life of the nation, the homes of ten millions of people, than Great Britain has done for a remote colony and a dependency, which she seems sometimes not very reluctant to have detached as an incumbrance.

Great Britain has constructed the Canadian canals to secure distant and sparsely settled provinces, whose commerce is small compared with ours, and upon which, therefore, the injury we could inflict in case of war, would be trifling as compared with that to which our Lake towns and commerce would be exposed in case England should obtain supremacy on the Lakes.

The above are some of the reasons why we most fully concur in the resolution unanimously adopted by the Convention by which we were appointed, "that canals, with dimensions sufficient to pass gun-boats from the Mississippi to Lake Michigan, and from the Atlantic to and from the Great Lakes, *will furnish the cheapest and most efficient means of protecting the Northern frontier.*"

2. Commercial Importance of the Ship-canals.

Your memorialists, having presented their views of the military importance of these canals, ask attention to some considerations showing that they have become a *commercial necessity.*

The configuration of the North American continent presents the most remarkable adaptation to internal commerce of any portion of the globe.

The great interior basin drained by the Mississippi and its tributaries, with ten thousand miles of steamboat navigation; the Lakes, with their shore lines of five thousand miles, and with more than ninety thousand square miles of surface, these great Mediterranean seas of the New World, can be connected with the great river of the West by a steamboat and ship-canal only thirty-six miles long!

The commerce of these Lakes, carried in fleets composed of sixteen hundred and forty-three vessels and steamers, reaches in value between four and five hundred millions of dollars per annum. The commerce of the Mississippi and its great tributaries, before the rebellion, it is believed, was not less in value. A direct union between these waters will be like the union of two oceans. The Suez canal does not compare in importance with a ship-canal from the Mississippi to the Lakes. No day should delay its accomplishment.

The outlet to the Atlantic by the East is equally remarkable with that of the South, and equally favorable to the commercial development and unity of our country. The arm of Almighty God cut down the barriers of the Alleghanies, and ordained that the ocean tides should flow through the highland passes of these mountains. The broad Hudson stretching away northerly towards the Lakes pointed to the sagacious statesmen of New York the pathway to empire. The genius of De Witt Clinton, quick to catch the clear intimation, consummated what nature had so nearly completed, and opened the way by the New York canals from the Atlantic to the Lakes. Illinois, by the aid of the Federal Government, followed, completing the water-channel from the Atlantic to the Mississippi; and now we have only to follow the finger of God, as interpreted by Clinton, and consummate what is so nearly done, and we have an east and west Mississippi from the Missouri to the Atlantic.

WE HAVE OUTGROWN OUR CANALS.

The great commercial fact of to-day, felt and realized, is, that we have outgrown our canals. The country is too big for them. The problem is, shall production stop its increase, or shall our canals be enlarged? The necessity of this enlargement is manifest by the enormous profits of the great railways, and the extravagant rates of transportation, showing that the quantity to be carried forward is so vast that carriers command their own terms. The warehouses and mammoth elevators of the Lake towns for the last two years have been crushed with freight; every thing which could be made to float on the Lakes and canals, has been taxed to the utmost, and proved insufficient to carry to market the products of the West. This necessity for greater facilities, and the failure in Congress of the bills for enlarging the New York and Illinois canals, have led to a zealous effort on the part of the West to obtain, by Canadian canals, that relief which is (we trust only temporarily) denied through our own country, and by our own Government. Illinois and Wisconsin, through their State authorities, and the Boards of Trade of several Lake cities, appointed delegates to Canada, to obtain, if possible, avenues to market for the vast accumulation of Western produce.

Necessity will force the West into new avenues to the Atlantic, unless the present are enlarged. That both Canada and Great Britain appreciate the value of this Western trade, is shown by their construction, for the purpose of securing it, of the Victoria Bridge at Montreal, at a cost of seven millions of dollars, and the Grand Trunk Railway at a cost of about sixty millions of dollars, in addition to the canals before referred to.

It is obvious from these and other facts, that we have reached that point, when, with our present means of transportation, the production of corn and other cereals cannot to any great extent be profitably increased. This condition should not surprise us. The canals were constructed while the West was in embryo. In 1837, the number of tons transported from west of Buffalo, on the Erie Canal, was 56,255. In 1861, the number reached 2,156,426.

The product of wheat and corn carried on the New York canals from the Lake States of Ohio, Michigan, Illinois, Indiana and Wisconsin, in 1850, was 252,000,000 bushels; in 1860, 354,000,000 bushels.

The population of these States, and Iowa, Kansas, Minnesota, Missouri and Nebraska, in 1850, was 5,403,595; in 1860, it was 9,092,009.

The value of Western products has increased more than 100 per cent. in the last four years. In 1859, it was, in round numbers, $53,000,000, and in 1862, $111,000,000. Our foreign exports are made up largely of breadstuffs and provisions. In four years they increased from $38,305,991 in 1859, to $122,650,043.27 in 1862, increasing in two years 180 per cent., and in three years 220 per cent. The amount of $122,650,043 for 1862, is exclusive of $11,100,043 which went out through Canada, making the aggregate over $133,750,000.

The following tables, compiled from the preliminary report of the census for 1860, will show the progress of the West, and will furnish data by which its present and future necessities may be more fully realized.

POPULATION OF THE NORTH-WESTERN STATES IN 1850 AND 1860.

STATES.	1850.	1860.	Ratio of Increase.
Illinois	851,470	1,711,951	101.05
Indiana	988,416	1,350,428	36.63
Iowa	192,214	674,913	251.12
Kansas	......	107,206	......
Michigan	397,654	749,113	88.38
Minnesota	6,077	172,123	2,732.36
Missouri	682,044	1,182,012	73.30
Ohio	1,980,329	2,339,511	18.13
Wisconsin	305,391	775,881	154.06
TOTALS	5,403,595	9,063,138	
Nebraska	....	28,841	......
GRAND TOTALS	5,403,595	9,091,979	68.25

VALUE OF REAL AND PERSONAL PROPERTY IN THE NORTH-WESTERN STATES IN 1850 AND 1860.

STATES.	1850.	1860.	Ratio of Increase.
Illinois	$156,265,006	$871,860,282	457.93
Indiana	202,650,264	528,835,371	160.95
Iowa	23,714,638	247,338,265	942.97
Kansas	....	31,327,895	......
Michigan	59,787,255	257,163,983	330.13
Minnesota	....	52,294,413	......
Missouri	137,247,707	501,214,398	265.18
Ohio	504,726,120	1,193,898,422	136.54
Wisconsin	42,056,595	273,671,668	550.72
TOTALS	$1,126,447,585	$3,957,604,697	
Nebraska	....	9,131,056	......
GRAND TOTALS	$1,126,447,585	$3,966,735,753	243.23

NOTE.—In the official reports at hand, no separation is made of the respective amounts of real estate and personal property in 1850.

NUMBER OF ACRES AND VALUE OF IMPROVED LANDS, AND RATIO TO TOTAL AREA IN THE NORTH-WESTERN STATES, IN 1850 AND 1860.

STATES.	TOTAL AREA. Acres.	IMPROVED LANDS, 1850.		Ratio of Improv'd Lands to Total Area.	IMPROVED LANDS, 1860.		Ratio of Improv'd Lands to Total Area.
		No. of Acres.	Value.		No. of Acres.	Value.	
Illinois.........	35,459,200	5,039,545	$96,133,290	14.21	13,251,473	$432,531,072	37.51
Indiana.........	21,687,760	5,046,543	136,385,173	23.26	8,161,717	344,902,776	37.71
Iowa...........	32,584,960	824,682	16,657,567	2.53	3,780,253	118,741,405	11.60
Kansas.........	73,470,720	....	....	.....	372,835	11,394,184	0.50
Michigan.......	35,995,520	1,929,110	51,872,446	5.35	3,419,861	163,279,087	9 77
Minnesota......	106,256,000	5,035	161,948	0.04	554,397	19,070,737	0.52
Missouri........	43,123,200	2,938,425	63,225,543	6.81	6,246,871	230,632,126	1.44
Ohio............	25,576,960	9,851,493	358,758,603	38.51	12,665,587	666,564,171	49.51
Wisconsin......	34,511,360	1,045,499	28,528,563	3.02	3,746,036	131,117,082	10.85
TOTALS.......	408,615,580	26,680,332	$751,723,133	6.52	52,199,030	2,118,232,640	12.27
Nebraska.......	214,964,480	....			122,582	3,916,002	0.05
GRAND TOTALS..	623,580,060	26,680,332	$751,723,133	4.27	52,321,612	2,122,148,642	8.13

QUANTITY OF WHEAT AND INDIAN CORN IN THE NORTH-WESTERN STATES IN 1850 AND 1860.

STATES.	WHEAT. Bush.		INDIAN CORN. Bush.	
	1850.	1860.	1850.	1860.
Illinois..............	9,414,575	24,159,500	57,646,984	115,296,779
Indiana..............	6,214,458	15,219,120	52,964,363	69,641,591
Iowa................	1,530,581	8,433,205	8,656,799	41,116,994
Kansas..............	....	168,527	....	5,678,834
Michigan............	4,925,889	8,313,185	5,641,420	12,152,110
Minnesota............	1,401	2,195,812	16,725	2,987,570
Missouri.............	2,981,652	4,227,586	36,214,537	72,892,157
Ohio................	14,487,351	14,532,570	59,078,695	70,637,140
Wisconsin............	4,286,131	15,812,625	1,988,979	7,565,290
TOTALS............	43,842,038	93,062,130	222,208,502	397,968,465
Nebraska............	....	72,268	....	1,846,785
GRAND TOTALS.....	43,842,038	93,134,398	222,208,502	399,815,240

NUMBER OF MILES OF RAILROAD IN OPERATION IN THE NORTH-WESTERN STATES IN 1850 AND 1860.

STATES.	1850.	1860.
Illinois	110.50	2,867.90
Indiana	228.00	2,125.90
Iowa	......	679.67
Kansas	......	....
Michigan	342.00	799.30
Minnesota	......	....
Missouri	......	817.45
Ohio	575.27	2,900.75
Wisconsin	20.00	922.61
TOTALS	1,275.77	11,113.58

With the canals enlarged as proposed, production may be stimulated a hundred fold, and yet still yield a fair profit to the producer. These enlarged canals, reducing materially the cost of transportation, will enable us to compete successfully in, and perhaps control the foreign market, for breadstuffs and provisions. Every acre of land west of the Lakes to the Rocky Mountains, and from Cairo and even Memphis north to, and including Minnesota, will be brought practically hundreds of miles nearer market, and of course every acre of land throughout this vast area will be increased in value.

This will stimulate emigration, settlement and production, and secure the early cultivation of the fertile lands of the Mississippi valley; and secure to our agriculturists the markets of the world. With these canals, the Western farmer can compete successfully with the grain-producing countries of the old world, and drive them from the field of competition.

It should be remembered that the increased production of food in Europe is limited by physical difficulties. The country is old, thickly peopled, and the good land is all improved. Mountains, barren wastes, and irreclaimable marshes, offer obstacles to any great increase in the production of food.

With us it is otherwise. We have a soil of inexhaustible fertility, a large portion of it as yet unbroken. There is spread out between the Lakes and the base of the Rocky Mountains, millions and millions of acres of the richest land on earth. This soil has a peculiarity of great significance. It is so admirably adapted to the use of labor-saving machinery, that although the North-West has sent not less than half a million of her most efficient laborers to the camp as volunteers, their absence has been so successfully supplied by labor-saving machines, that the quantity of land cultivated has not been lessened, nor the crops materially diminished. God has so fashioned this land, that with small labor it will yield the most bountiful return in endless crops of food. He has fitted it to be the garden and granary of the world, richer even than the Valley of the Nile. He causes the sun to shine, and the rain to fall upon this land, and clothes it with a rank and luxurious vegetation which annually decays where it grows, or feeds the prairie fires which sweep over it in autumn. We have the land, the laborer is ready, but without these enlarged canals, the labor will not be remunerative, and the land will not be cultivated.

Corn, for want of adequate means of transportation, is, on the Western prairies, annually consumed for fuel. This does not pay. Shall Europe starve for bread, and our rich prairies remain uncultivated for want of these canals to carry the products to market? Let Congress answer by its action on these great questions now presented.

By reference to the reports of the Boards of Trade and Mercantile Associations submitted to the Convention, and which we believe to be entirely reliable, it appears that the enlarged canals would reduce the cost of transportation between Chicago and New York at least ten cents, and between the Mississippi and the Atlantic at least fifteen cents a bushel. Divide this saving between the Western producer and the Eastern consumer, and while you raise the price of every bushel of wheat and corn to the farmer, you reduce the price of every loaf of bread in every house in New England and the sea-board cities. The crop of 1862 shipped to the East through the canals alone, exceeded one hundred millions of bushels! When we remember that the West pays annually more than fifty millions of dollars for transporting its produce to market, it is obvious that there would be saved on the transportation of a single crop more than the entire cost of these improvements.

But it is not the trade of the Mississippi Valley only, vast as it now is, and almost incalculable in its future, that will require these enlarged canals. All this will at no distant day be augmented by contributions from the auriferous regions of the Rocky Mountains, the Valley of the Columbia, and the Pacific Coast. The mineral wealth of this region being rapidly developed is not yet appreciated. The copper and iron of Lake Superior, the lead of Illinois and Wisconsin, the inexhaustible coal-fields of the great interior basin, and the silver and gold of the Rocky Mountains, added to the agricultural wealth of the great interior, make it among the most favored portions of the globe. To develop these advantages requires the immediate construction of these canals.

THE WORDS OF BENTON.

The great statesman of Missouri, THOMAS H. BENTON, a man whose vast information and ideas were worthy of the Mississippi valley, in 1847 addressed the River and Harbor Convention in words worthy of being recalled to the attention of the American people to-day.

He says:

"The lake and river navigation of the Great West, to promote which your Convention is called, very early had a share of my attention, and I never had a doubt of the constitutionality or expediency of bringing that navigation within the circle of internal improvements, by the Federal Government, when the object of the improvement should be of general and national importance.

"The junction of the two great systems of waters which occupy so much of our country, the Northern Lakes on the one hand, and the Mississippi River and its tributaries on the other, appeared to me to be an object of that character, and *Chicago the proper point for effecting the union;* and near thirty years ago I wrote and published articles in a St. Louis paper in favor of that object, indicated and almost accomplished by nature herself, and wanting from man little to complete it. These were probably the first formal communications upon authentic data in favor of the Chicago canal.

"The nationality of the Chicago canal and the harbor at its mouth are by no means new conceptions with me.

"The river navigation of the Great West is the most wonderful on the globe, and since the application of steam power to the propulsion of vessels, possesses the essential qualities of open navigation. Speed, distance, cheapness, magnitude of cargoes, are all there, and without the perils of the sea from storms and enemies. The steamboat is the ship of the river, and finds in the Mississippi and its tributaries the amplest theatre for the diffusion and the display of its power. Wonderful river! Connected with seas by the head and by the mouth, stretching its arms towards the Atlantic and the Pacific—lying in a valley which is a valley from the Gulf of Mexico to Hudson's Bay—drawing its first waters not from rugged mountains, but from the plateau of the Lakes in the centre of the continent, and in communication with the sources of the St. Lawrence and the streams which take their course north to Hudson's Bay—draining the largest extent of richest land, collecting the products of every clime, even the frigid, to bear the whole to market in the sunny South, and there to meet the products of the entire world. Such is the Mississippi. And who can calculate the aggregate of its advantages, and the magnitude of its future commercial results?"

Hear Silas Wright, as worthy to speak for the East, as Benton for the West:

"I am aware that questions of constitutional power have been raised in reference to appropriations of money by Congress, for the improvement of Lake harbors, and I am well convinced that honest men have sincerely entertained strong scruples upon this point; but all my observation and experience have induced me to believe that these scruples, where the individual admits the power to improve the Atlantic harbors, arises from the want of an acquaintance with the Lakes and the commerce upon them, and an inability to believe the facts in relation to that commerce, when truly stated. It is not easy for one familiar with the Lakes and the Lake commerce, to realize the degree of incredulity, as to the magnitude and importance of both, which is found in the minds of honest and well-informed men, residing in remote portions of the Union, and having no personal acquaintance with either; while I do not recollect an instance of a Member of Congress, who has traveled the Lakes and observed the commerce upon them within the last ten years, requiring any further evidence or argument to induce him to admit the constitutional power, and the propriety of appropriations for the Lake harbors, as much as for those of the Atlantic coast. I have long been of the opinion, therefore, that to impress the minds of the people of all portions of the Union with a realizing sense of the facts as they are in relation to these inland seas, and their already vast and increasing commerce, would be all that is required to secure such appropriations as the state of the National Treasury will from time to time permit, for the improvement of the Lake harbors."

But the scruples which Benton and Wright sought to remove, disappear in the light of the census returns showing the national, not to say continental, character of the commerce to be relieved: and now that the action of Great Britain has given to these improvements the character of necessary, defensive military works, these scruples disappear, and are no longer entitled to serious consideration. It is not too much to say that to the magnificent enterprise of the Erie Canal, the North-West owes its existence. This great section would have been as yet in its feeble infancy, but for the enterprise of Clinton and the genius of Fulton.

The enlargement of the canals between the Mississippi and the Atlantic would create a new era, from which would date another career of advance and progress equally rapid and important. The wave of emigration, checked by the war, is already returning, and will soon be upon us with increased volume.

The expenditures asked for by the contemplated improvements are light indeed: so much has already been done by Nature and by the States through which the improvements are to pass, that the cost of the completion of the ship-canal will be small compared with the results. We believe the Illinois canal can be constructed in the manner proposed by the Legislature of Illinois to the last

Congress, without costing the National Treasury a single dollar, that its tolls would soon pay interest and principal upon its cost, and thus this great national work, free at all times for the military purposes of the Government, would soon, having paid its cost, become free to the vast and constantly increasing commerce of the Lakes and the Mississippi.

These enlarged communications between the Mississippi and the Lakes and the Atlantic would save, every year, in lessening the amount paid for transportation, more than their cost. There is not an acre of land between Lakes Michigan and Superior and the western barrier of the Rocky Mountains, and including the Valley of the Mississippi to New Orleans, but would be increased in value, and the aggregate of such increase would bear no proportion to the amount required to complete the works.

There is not a bushel of wheat or corn, nor a barrel of pork or of beef, nor of any article of food in this whole area, but would be enhanced in value. The year these works should be completed they would add to the taxable property of the nation, an amount the taxes upon which in a single year would pay off their cost.

Such we believe, without exaggeration, are some of the advantages commercially and economically to result from these improvements.

We have referred to the commerce and trade of the Lakes and the Mississippi combined as vastly greater than our foreign commerce, and as supplying the bulk of our foreign exports. This is the West of to-day, with less than one-twentieth part of its available land improved. Stimulate industry, invite emigration and improvement by these canals, and who can estimate its future? What figures or language shall describe its greatness?

To render complete this great national work it will be necessary to clear the rapids of the Mississippi river at two points, namely, above Keokuk, and above Rock Island; this is an improvement long demanded by every State bordering on the great river, and of such acknowledged importance as already to have been partially accomplished by the Federal Government.

This will perfect the water-communication both between the extreme Northern States and the South, and the same States and the East; the first by the river alone, the latter by river, canals and lakes combined.

The outlay required for the removal of obstructions at these two points, to make good the navigation, is inconsiderable, and the advantage most important in a military as well as a commercial point of view.

3. National Unity will be forever secured by these Canals.

No reflecting mind who has marked the events of the last two years, but will admit that among the influences that have made separation and disunion impossible, was the Mississippi river. The great river of the West has been strong enough to hold the Union together. Never in the darkest, gloomiest hour of the rebellion, has the West considered it a debatable question that she could ever, under any circumstances, consent to separation. Her gallant soldiers have marched right on from Cairo to the Gulf, like the current of her great river, resistless, overcoming every difficulty, triumphing over every obstacle, until no rebel flag now floats upon her waters. She was deaf to the overtures of the traitors, who sought by alluring promises of commercial advantages to seduce the North-West from her fealty to the Nation. The West means to maintain the unity and integrity of the whole country. With one hand she grasps the South, and with the other she clasps the East, and she will never consent to reach the ocean in either direction through foreign territory.

But it must have occurred to every thoughtful mind how the ties which bind us together would be strengthened and multiplied by these ship-canals, creating another Mississippi from St. Louis, and Kansas, and St. Paul, to New York and Boston. It has been well said, that the myriad-fibered cordage of commercial relations, slight in any individual instance, but indissoluble in their multitudinous combination, produces such unity of purpose, unity of interest, intelligence, sentiment, and national pride, and social feeling, and that homogeneousness of population which unites peoples and maintains nationalities. Such will grow up with a power which no sectional feeling can break between the East and West, when connected together by these canals.

It is a curious fact in our history that the same man who was the father of nullification, the author of the secession heresy—the man who planted the seeds of this bloody rebellion, and nurtured them while he lived—in his earlier and better days was a truly national statesman, with an enlightened patriotism which embraced the whole country.

In 1824, John C. Calhoun, then Secretary of War, in advocating the construction of roads and canals by the National Government, said: "Let us bind the Republic together, let us conquer space, by a perfect system of roads and canals."

It was said by Montesquieu, that a Republic could not exist and govern a large territory. There was some truth in the remark when he made it, and it has in it still enough of reason, in spite of steam, railways and telegraphs, and other agencies that annihilate distance, to make it wise for our statesmen to bind our different sections together by every means in their power. The rebellion has demonstrated the necessity of a closer union and a more consolidated nationality. No agency will be more effective in securing these, than these great ship-canals.

The nation has expended its millions of treasure without regard to the amount, and its blood has been poured out like water to open the Mississippi, and yet no one has been found to declare that the cost has been too great for the object. Such is the profound conviction that we *must* maintain the integrity of the Union and a free passage to the Gulf and the Sea. The Eastern pathway to the Ocean by these enlarged canals would be still more important, and would serve still more strongly to bind the Union together. And yet this can all be secured by a sum less than a month's military expenditure in the valley of the Mississippi, and without one drop of precious blood.

If the map of the territory, which is to be connected by these canals, extending from the West to the shores of the Atlantic, were laid over the map of Europe, that portion of the globe which for the last thousand years has engrossed the attention of the civilized world, would be entirely covered. It would overspread monarchies, empires and nationalities, which for ages have been antagonistic, belligerent—the great battle-fields of Europe. It would cover the theatre of the great wars, which have desolated and depopulated again and again that continent, from France and Waterloo to Sebastopol. Human beings by the million have been sacrificed in the wars of the Fredericks, of the Louises, of the Phillips, and the Charleses, of the Marlboroughs and of the Buonapartes. Millions and millions of treasure wrung from the toil of the laboring masses, have been expended in fortifying frontiers, and the operations of these wars. Rivers of blood have flowed, so that you cannot take a day's ride in Europe, without passing over fields memorable for human slaughter. Shall these scenes of butchery and desolation be re-enacted in our own beloved country? Shall this fair country, lately so peaceful, prosperous and happy, break into fragments? Shall the Hudson, the Susquehannah, the Delaware, and the Ohio bristle with fortifications? Shall the Atlantic States contend in battle with the generous West? Shall we ever re-enact upon these fair prairies and broad lakes the bloody pages of European history? Shall fratricidal wars, with all their horrors, their merciless expenditures of blood and treasure, darken the future pages of American history?

God forbid. Could some divine agency, a thousand years ago, have made of Europe a great confederate nationality, levelling its dividing mountains, and mingling its clans into one great homogeneous people, and made it free, virtuous, and wise enough to be united, what untold misery and suffering would have been prevented.

No levelling of dividing mountains is here necessary. *God* in his goodness has fashioned our country, vast as it is, for unity. He has given us one language, the same laws, and one glorious flag. He has made one great nationality a necessity. He has blessed us with liberty. Let us second God's plans, and aid and strengthen by every generous means the influences which shall hold us together forever.

What are a few millions expended, if the tendency is to strengthen the bonds of our Union?

Could the present horrid rebellion have been prevented at any cost of money, or treasure, how wise, how economical, how beneficent the expenditure?

Let no narrow jealousies, let no sectional prejudices delay these great improvements so important to the general welfare, so necessary to our security, so favorable to our commercial development, so just to the Western producer, so beneficial to the Eastern consumer, so essential to the growth of the West; but above all, so indispensably useful in binding our wide territory in one perpetual Union.

It seems to your memorialists that no one can study the outlines of our country without becoming satisfied that the works contemplated are not only necessary, but *inevitable*. Such vast advantages at such small cost will not be neglected by a people so sagacious and enterprising as those represented by the American Congress.

Let us then crown the mighty military struggle in which we are engaged and which now seems to be approaching a triumphant close, let us crown it by one of those signal triumphs of peace, which, not less than the victories of war, shall exhibit our devotion to the Union, and our determination that by its multiplied blessings we will make it perpetual.

Then, looking down the future, may we contemplate an ocean-bound Republic—a glorious band of an unbroken brotherhood of States—with its hundreds of millions of people, speaking the same language, living under the same laws, worshipping the same God, and over all, floating the same flag, consecrated forever to Liberty and Union.

ISAAC N. ARNOLD,	Illinois.
A. A. LOW,	New York.
P. CHAMBERLIN,	Ohio.
JAMES A. McDOUGALL,	California.
EZRA NYE,	New Jersey.
JUSTIN S. MORRILL,	Vermont.
HENRY L. DAWES,	Massachusetts.
GEORGE W. JULIAN,	Indiana.
THOMAS M. EDWARDS,	New Hampshire.
RUSSELL BLAKELEY,	Minnesota.
D. R. ANTHONY,	Kansas.
T. C. HERSEY,	Maine.
M. M. DAVIS,	Wisconsin.
SAMUEL L. CASEY,	Kentucky.

INDEX.

www.ingramcontent.com/pod-product-compliance
Lightning Source LLC
LaVergne TN
LVHW050513100826
845148LV00002B/327

* 9 7 8 1 4 2 5 5 2 1 3 5 6 *